OECD Green Growth Studies

Fostering Green Growth in Agriculture

THE ROLE OF TRAINING, ADVISORY SERVICES AND EXTENSION INITIATIVES

Please cite this publication as:
OECD (2015), *Fostering Green Growth in Agriculture: The Role of Training, Advisory Services and Extension Initiatives*, OECD Green Growth Studies, OECD Publishing, Paris.
http://dx.doi.org/10.1787/9789264232198-en

ISBN 978-92-64-23215-0 (print)
ISBN 978-92-64-23219-8 (PDF)

Series: OECD Green Growth Studies
ISSN 2222-9515 (print)
ISSN 2222-9523 (online)

Foreword

Investing in knowledge to support the adoption of environmentally-friendly farm practices is commonly perceived as a key driver behind innovation processes in agriculture. Yet changes at the national and global levels have led to dramatic changes in the orientation of advisory services, how these are organised, and their methods of intervention. This report examines the role, performance and impact of farm advisory services, as well as the training and extension initiatives undertaken in the OECD area to foster green growth in agriculture. The merits of the different types of providers are also discussed and the experience of selected OECD countries presented.

Assessing the impact of agricultural advisory services, training and extension measures on green growth involves a range of methodological issues, but for which evaluations of outcomes and assessment of their overall cost-effectiveness is scarce. Nevertheless, a key conclusion of this report is that there is no one-size-fits-all evaluation methodology and that any evaluation of the impact of these measures should take into account all actors that provide agricultural advisory services, training, and extension measures as they are part of a wider agricultural knowledge and innovation system in which multiple stakeholders interact.

This report contributes to OECD work on green growth which emphasises the importance of research, development, innovation, education, extension services and information to increase productivity in a sustainable way. This report was prepared by the OECD's Trade and Agriculture Directorate and was declassified by the OECD Joint Working Party on Agriculture and the Environment in January 2015.

Dimitris Diakosavvas was project leader and is the principal author of this report. Chapter 5 draws on background papers prepared by consultants for the five case studies: Bruce Kefford and Clive Noble (Australia); Rivellie Tschuisseu and Pierre Labarthe (Canada), Janet Dwyer and Matt Reed (England and Wales), Dimitris Damianos (Greece) and Brian Bell and Michael Yap (New Zealand). A further paper prepared by Clunie Keenleyside also contributed to the present report. Comments and review from OECD colleagues are also appreciated and acknowledged, including Nathalie Girouard, Justine Garrett and Annabelle Mourougane. Françoise Bénicourt and Theresa Poincet provided invaluable secretarial assistance throughout the production process. The report was prepared for publication by Michèle Patterson, who also co-ordinated its production.

Table of contents

Tables

Figures

Executive summary

Agricultural advisory services, training and extension initiatives play an important role in supporting green growth in agriculture and enabling farmers to meet new challenges, such as adopting environmentally sustainable farming practices and improving their competitiveness. At present, there is renewed interest in agricultural advisory services in many countries.

This report examines investment in these initiatives in the OECD area and discusses the use and merits of different types of providers, both alone and in combination. It provides guidelines and recommendations on best practice, summarising the features of successful advisory, training and extension measures that foster green growth. It then takes a closer look at investments in knowledge to support environmentally-friendly farming practices in a selection of OECD countries, namely Australia, Canada, England and Wales, Greece and New Zealand.

Key findings

In the context of green growth, advisory, training and extension services have two main roles: i) to incentivise farmers by making them aware of the possible benefits of different measures, while retaining or improving the economic performance of the farm (economic role); and ii) to encourage and facilitate the adoption of appropriate agri-environmental land management practices and therefore maximise environmental benefits (environmental role). These services take a range of forms within the OECD area, in terms of their delivery and funding methods, scope and organisational structures.

The number of potential providers of support for environmental management on the farm is also considerable, and includes: government advisory and extension services (agricultural and environmental); agents and advisers employed by individual farmers; farmers' associations and producer groups; specialist environmental service providers, including non-governmental organisations; agencies involved in the delivery and control (compliance-checking) of agri-environmental interventions supported by advisory measures; and co-operatives and informal self-help and peer groups created by beneficiaries.

Diverse institutional options exist to finance and provide agricultural advisory services. Since all options have advantages and disadvantages, it is an important task when evaluating advisory service policies to identify the mix of options best suited to support a country's agricultural development strategy and farm diversity in the most cost-effective way by taking country-specific conditions into account.

Shifts in the orientation, organisation and methods of providing agricultural advice, training and extension services have brought both gains and losses. On the one hand, intensified transactions between farmers and providers have led to the development of a system of sharing and exchange of relevant knowledge, which is beginning to spread to the research community. On the other hand, there are concerns – particularly about availability and access to these services – expressed by certain groups of farmers, particularly those who own small-size farms. Knowledge gaps on certain issues nevertheless continue to exist, e.g. on the effectiveness of agro-ecological practices.

Despite growing awareness in OECD countries of the value of these services, there has been little evaluation of their outcomes, overall effectiveness and efficiency. Available evaluation studies are largely qualitative, mainly focused on "snapshot" evidence and often based on small numbers of participants, in interviews and surveys.

The limitations of the existing research evidence indicate that the effectiveness of different forms of training, advice and extension is not comprehensively understood. Public and private investment decisions are therefore being made on a narrow evidence base, suggesting they may not be providing optimal returns.

Assessment can be problematic. The main challenges derive from problems such as multiple goals, attribution problems, lagged effects, spillover effects, data problems, sample attrition, and difficulties to establish a baseline. Because many factors affect the performance of agriculture in complex and sometimes contradictory ways, it is difficult to quantify the relationship between advisory services, training and extension and their impact at the farm level. The decision-making processes of farmers are highly contingent, increasing the difficulty of assessing how advisory services, training and extension impact on farming decisions.

Given these methodological difficulties, together with the great diversity of approaches used by OECD countries, an accurate assessment of the overall impact of agricultural advisory, training and extension measures in fostering green growth is not feasible. Many situation-specific factors, including eligibility criteria for obtaining advisory services impinge on the effectiveness of advisory, training and extension measures. Nevertheless, evidence from the Australian, England and Wales, and New Zealand case studies suggests that these measures constitute a vital element in supporting the transition towards sustainable agriculture: they lead to increases in investment returns and gains in productivity, and improvement of the environment is significant, although more difficult to quantify.

Some preliminary policy recommendations

- Advisory services, training and extension measures should be targeted and have clear objectives regarding their role within the policy mix.
- The key ingredients for persuading farmers and enabling farmers to adopt green growth practices are credible, relevant and up-to-date business-acumen advice, training and extension.
- Both public and private funding of initiatives have a role to play and will reflect government policies and resources, the nature of the issues, the type of provider, and the purpose of the measure.
- Agencies that deliver advice, training and extension services to support agri-environmental management will need to be well co-ordinated, effective in reaching different groups of farms and types of farming, and capable of delivering a full range of services.
- There are no general one-size-fits-all evaluation methodologies and approaches. Evaluation of impacts should take into account all actors involved in their provision.

Chapter 1

Assessment and recommendations

This chapter proposes guidelines and lessons for best practice to improve the effectiveness of advisory, training and extension measures to achieve green growth objectives in the agricultural area.

How can advisory, training and extension measures be used most effectively in the future?

This report illustrates the diversity of advisory services, training and extension measures in nurturing agri-environmental management used today in OECD countries. It shows how the objectives, methods, delivery agencies and sources of funding have changed, and continue to change. Two broad trends have been identified, with both marking a shift from relative simplicity to increased sophistication.

The first trend is the demand by society on farmers and land managers to deliver a broad range of environmental public goods while continuing to produce food, feed and other marketable goods. The second is in the funding and delivery of advisory services, training and extension measures, combined with the decline of public sector dominance (more marked in some countries than others) and the emergence of a wide range of actors with differing sets of skills and priorities.

It can be argued that these changes have brought with both gains and losses. On the positive side, intensified transactions between farmers and providers have led to the development of a system of sharing and exchanging relevant knowledge, and which is beginning to include the research community. Many more farmers now have an opportunity to choose the type and provider of support services that best suit their needs, especially if they pay for these services. Where public funding remains, it is no longer necessarily linked to public provision, but instead may harness the strengths of other providers, such as NGOs, with specialist skills and peer groups that engage directly with the farmers.

There are nevertheless concerns, particularly about certain groups of farmers and particular types of support. Empirical studies found that in several OECD countries specific groups of farmers lack access to support services. The Greek case study, for example, reports that only about 0.3% of farmers are eligible to receive advice under the EU FAS system. Moreover, there also concerns about the impartiality and reliability of the advice provided (Damianos, 2015).

If the environmental potential of agriculture is to be fully realised, the varied and pluralistic mix of agencies that deliver advice, training and extension services to support agri-environmental management will need to be well co-ordinated, more effective in reaching different groups and types of farmers, and capable of delivering a full range of services.

A different problem may arise where farmers rely for agricultural advice on private sector or peer group providers, whose priority will be to offer support that matches farmers' perceived needs and priorities. This form of support is likely to focus on farm business performance and may fail to deliver support for agri-environmental management beyond basic regulatory compliance. This raises an important question about the need for continued public funding of some types of agri-environmental support, especially for those countries that have moved towards more paid-agricultural advisory services.

A horizon-scanning report on the future of the global food system highlighted the need for improvements in extension and advisory services in high- and middle-income countries (Foresight, 2011). It is not possible to estimate current public expenditure on advisory services, training and extension measures to support agri-environmental management, but where figures are available the estimates of annual public expenditure are significant, although they remain a very small proportion of total public support for agriculture.[1]

In the **United States**, agri-environmental technical support remains largely a public service delivered by the Department of Agriculture, mainly within the context of government conservation programmes. Some observers see the lack of technical assistance as a major barrier to the adoption of conservation practices and uptake of these programmes, while producers, ranchers, environmentalists, and wildlife advocates continue to raise the issue of technical assistance and the need for additional support. Recent Farm Bills have repeatedly added natural resource concerns to the conservation mission, leaving many to question whether the current delivery system for technical assistance can function effectively (Stubbs, 2010).

At a time of increasing pressure to reduce public expenditure on agriculture, it is important to be clear about the overall objectives and the role of advisory services, training and extension measures within the policy mix. This will inform decisions on the type and quality of the provisions required and where to strike the balance of funding between these measures and financial support for appropriate land management. Some observers in the United States believe that the technical assistance element should account for at least 30% of the combined expenditure on technical and financial support for conservation.

As one group has argued, "technical assistance multiplies the benefits of financial assistance, and financial assistance multiplies the benefits of technical assistance. Sometimes, technical assistance alone is enough. Sometimes technical assistance needs to be coupled with small and perhaps short-term incentive payments. In other cases, no change can occur without substantial financial assistance. The key is to get the right mix." (SWCS, 2007)

The way in which advisory services, training and extension measures are designed, delivered and funded depends on their objectives. Defining these objectives will help clarify the need for public spending in this area as well as establish what elements need to be focused on In practice, several objectives may be pursued at once, with trade-offs required. In some cases the objective will be to secure compliance with environmental regulations, or to improve the outcome of conservation incentive schemes, or to ensure access to support for target groups in the absence of public funding.

In light of these objectives, the scope and depth of support required can be calibrated according to the core mission, which may range from a simple need to make farmers aware of an environmental issue, to the provision of detailed technical support or training at the individual farm level. The main focus of the support may vary from achieving minor modifications of conventional agricultural production, to bringing about major changes in environmental management that will have significant consequences for farming systems.

Although social theories can explain how some advisory services, training and extension measures work, it would be a mistake to assume that conclusions on the efficacy of "traditional" extension services that promote increased productivity and income can simply be extrapolated to the new agri-environmental support measures. These latter measures are different in several respects.

The limited quantitative research that has been done shows mixed results, and therefore only general guidance can be given on designing and delivering effective training and extension services to support agri-environmental policy implementation within the OECD. Based on the examples given in this report, and provided by other studies, successful advisory, training and extension service measures to foster green growth seem to share some characteristics.

Since all options have advantages and disadvantages, it is an important to identify the mix of options best suited to support in a cost-effective way a country's agricultural development strategy.

Characteristics of successful advisory, training and extension measures to foster green growth

Clearly define the desired environmental and socio-economic objectives and outcomes from the outset

Objectives and desired outcomes might be specified in terms of expected changes in farm management, farmers' understanding, behaviour and attitudes, and even farm income. Only after the relevant outcomes are specified is it possible to identify the target land, farming systems, type of farm and farmer able to deliver the objectives. It is important to stress that even where objectives are similar, measures should be adapted to suit different types of users and local circumstances.

Demonstrate good understanding of the concerns and priorities of the targeted farmers

A key element of the design process is an understanding of the main agricultural management concerns and priorities of the target farmers and of how they will interpret the environmental objectives. Without this, it is very difficult to design advisory services, training and extension measures that will be used effectively by farmers. There should be no unfounded assumptions about these priorities and concerns, which may be complex and sometimes unexpected. The active participation of the intended "consumer", i.e. farmers, in the design process overcomes these problems and can help to improve uptake, as effective knowledge transfer is a two-way exchange.

Targeting a broad range of users by employing a variety of approaches

Advice, training and extension should target the largest possible range of farmers and land managers, using a variety of approaches and a combination of different mechanisms. Market segmentation is highly desirable because farmers are a very heterogeneous group. Characteristics such as age and farm structure influence the uptake of these measures.

Emphasise positive business opportunities

From the point of view of green growth, a successful advisory, training and extension approach needs to lead to the adoption of farm practices that are simultaneously profitable and beneficial to the environment. When transferring knowledge it is important to emphasise the potential business opportunities for farmers that the adoption of such practices can create. In **England**, for example, a survey by Farming Futures found that 53% of farmers recognised that adopting measures to tackle climate change offered business opportunities, and that adopting these measures could improve the profitability of their farms (Farming Futures, 2010).

Use trusted providers to give credible advice and information

Advisory, training and extension measures do not have automatic legitimacy and credibility – these must be earned. There is much research indicating that the qualities and credentials of the person providing information is at least as important to uptake as the scientific quality of that information. The key determinant of an adviser's credibility is trust (DEFRA, 2013; 2010). Farmers are more likely to listen to other farmers, or to people whom they know well and trust. Adviser credibility and trust can be easily lost by the support of an innovation or practice clearly unsuited to local circumstances.

Make ease of use a priority; one size does not fit all

Several different types of support and methods of communication are offered to farmers, recognising that *one size does not fit all* and *ease of access* are essential. This may be physical access (a helpline, printed publications, and one-to-one advice in a local office), or technical access (to broadband services and an interactive website with detailed information and maps), or via existing providers of other services that are already trusted by the recipient. Opportunities for peer group support, networking and co-operation are used to the full (for example by offering facilitation) where this would improve the effectiveness of advisory, training and extension measures.

Upgrade advisory skills to meet emerging challenges of increasing productivity in a sustainable manner

Technical staff who are trained in environmental and agricultural management, farm business development and extension techniques, and have the capacity to gain the confidence of farmers for the support they provide are a prerequisite in order to succeed in raising awareness and changing the perceptions of farmers to accelerate their adoption of farm practices that foster green growth.

In the context of green growth, the tasks and responsibilities of extension service need to be broadly based and holistic in content and scope, thus no longer limited to overseeing agricultural technology transfers. Nor is it feasible, or desirable, to continue to expect advisers to be proficient in all areas of expertise. Extension agencies, services, and workers will need to exercise a more pro-active and participatory role, and serve as knowledge/information "brokers" who initiate and facilitate mutually meaningful and equitable knowledge-based transactions among agricultural researchers, trainers, and primary producers.

A greater emphasis needs to be placed on providing advisors with communication skills to enable them to influence attitudes that lead to changes in behaviour that have positive environmental impact. Advisers need to be familiar with the wide range of approaches and possible forms of communication with farmers in different circumstances; these include demonstrations, peer group learning, individual contacts on- and off-farm, formal training courses, and the use of IT methods.

Make effective links with professional and research bodies

Effective links between research organisations and professional bodies should be in place as the key knowledge source for advisers to maximise opportunities to exchange knowledge. There is a need for a rapid review process by the delivery body to evaluate progress against objectives and seek feedback to guide any adjustments needed (for example, to address requests for technical support not anticipated at the planning stage, or to re-focus efforts on certain groups of farmers).

Keep expectations realistic regarding impacts of advisory, training and extension measures

It must be recognised that where the advice provided is sufficient to induce a change in farmers' attitudes, this is not by itself sufficient to change behaviour. Even with the most expert and persuasive advice, training and extension services, farmers are not likely to change their management practices unless they can be convinced that the proposed changes are consistent with their goals. Therefore, expectations about final results must be realistic.

Promote "adoptable" farm practices

Adoption of farming practices, including those for the sustainable management of natural resources, is complex and multifaceted; it is nevertheless reasonably well studied (Pannel et al., 2006; Pannell, 2009; Lambert et al., 2006; Prokopy et al., 2008; Caswell et al., 2001; Rogers, 2003). The literature suggests that adoption of these practices by farmers and land holders can be explained in terms of characteristics of the learning process, the potential adopters, or by the conservation practices.

The literature on adoption of sustainable management of natural resources has long asserted that practices, including conservation practices, with the highest rates of adoption are those that were built on existing practices, and required the least amount of off-farm or scarce resources. Farmers are most likely to adopt technologies with certain characteristics. Favoured technologies are those that: i) have a relative advantage over other technologies (e.g. lower costs, higher yields); ii) are compatible with current management objectives and practices; iii) are easy to implement; iv) are capable of being observed or demonstrated; and v) are able to be adopted on an incremental or partial basis.

One implication is that if the farming practice being promoted is not sufficiently attractive and therefore not adopted, advisory services, training and extension initiatives would not succeed in bringing about positive environmental behaviour change. Providers of advisory, training and extension measures should, therefore, first assess whether a farming practice is adoptable before proceeding with the measures to promote their uptake. From the point of view of green growth, the real challenge is to identify or develop farm practices that are adoptable, beneficial to the environment and also economically superior to the practices they are supposed to replace.

Take into account the prevailing policy regime and other actors of the Agricultural and Knowledge System

The prevailing broader policy regime is one of the key drivers affecting the adoption of farm practices to support agri-environmental management and can variously facilitate or hinder the effectiveness of advisory, training and extension measures. A non-accommodated policy environment could have a high opportunity cost in terms of foregone benefits from these measures, creating a divergence between potential and actual benefits. Moreover, evaluation of the performance and impacts of these measures should consider that these measures are part of a wider agricultural knowledge and innovation system.

Agricultural advisors and other actors of such system – who can belong to public, private, the third sector and farmers' organisation – operate in a collective way, co-operating or competing.

Crucial to contemplate evidence-based policy-making

It is clear that more research is needed on the design, implementation and evaluation of the effectiveness of advisory, training and extension measures to support agri-environmental policy implementation in OECD countries where agricultural land is expected to provide a range of environmental services. This could usefully focus on more detailed evaluation of existing measures and initiatives, on identifying the need for public intervention and funding for these measures to support farmers and land managers in delivering environmental objectives, and on identifying the most cost-effective methods to ensure these measures are demand-driven and needs-based.

Note

1. For example, approximate annual expenditure of: around EUR 145 million on the FAS in the European Union; AUD 55 million on LandCare in Australia; and USD 700 million on the Conservation Technical Assistance Programme in the United States (own figures based, on EC (2010, Youl, Marriott and Nabben (2006) and Stubbs (2010) respectively).

Bibliography

Caswell, M., K. Fuglie et al. (2001), *Adoption of agricultural production practices: lessons learned from the U.S. Department of Agriculture Area Studies Project*, Agricultural Economic Report No. (AER792), Economic Research Service/USDA.

Damianos, D. (2015), "An Analysis of the Performance of Farm Advisory Services as related to Fostering Green Growth – The Case of Greece", Consultant Report, OECD internal document.

Department for Environment, Food and Rural Affairs (DEFRA) (2013), *Review of environmental advice, incentives and partnership approaches for the farming sector in England*, https://www.gov.uk/government/publications/review-of-environmental-advice-incentives-and-partnership-approaches-for-the-farming-sector-in-england.

DEFRA (2010), *Agricultural Advisory Services Analysis*, July, London.

European Commission (EC) (2010), On the application of the Farm Advisory System as defined in Article 12 and 13 of Council Regulation (EC) No. 73/2009, Report from the Commission to the European Parliament and the Council, COM(2010) 665 final, http://eur-lex.europa.eu/legal-content/EN/TXT/?uri=CELEX:52010DC0665.

Farming Futures (2010), http://www.farmingfutures.org.uk/about-farming-futures.

Foresight (2011), *The Future of Food and Farming: Final Project Report,* The Government Office for Science, London, https://www.gov.uk/government/collections/global-food-and-farming-futures.

Lambert, D., P. Sullivan, et al. (2006), "Conservation Compatible Practices and Programs: Who Participates?", USDA/ERS, *Economic Research Report* 14, www.ers.usda.gov/media/284696/err14_1_pdf.

Pannell, D. (2009), "Technology change as a policy response to promote changes in land management for environmental benefits", *Agricultural Economics*, Vol. 40, No. 1.

Pannell, D., G. Marshall et al. (2006), "Understanding and promoting adoption of conservation practices by rural landholders", *Extension*, Vol. 46, No. 11.

Prokopy, L., K. Floress, D. Baumgart-Getz and D. Klotthor-Weinkauf (2008), "Determinants of agricultural best management practice adoption: Evidence from the literature", *Journal of Soil and Water Conservation*, September/October, Vol. 63, No. 5.

Rogers, E.M. (2003), *Diffusion of Innovations*, Fifth edition, The Free Press, New York.

Stubbs, M. (2010), *Technical Assistance for Agriculture Conservation*, Congressional Research Service, http://www.crs.gov/.

Soil and Water Conservation Society and Environmental Defence (SWCS) (2007), *Technical Assistance for Farm Bill Conservation Programs: a report from the Soil and Water Conservation Society and Environmental Defence,* *http://www.swcs.org/documents/filelibrary/TechnicalAssistanceAssessment.pdf*.

Youl, R., S. Marriott and T. Nabben (2006), *Landcare in Australia Founded on Local Action*, SILC and Rob Youl Consulting Pty Ltd, http://www.agriculture.gov.au/ag-farm-food/natural-resources/landcare/publications/landcare-australia.

Chapter 2

The evolving role of advisory, training and extension measures

This chapter discusses the evolving role of advisory services, training and extension initiatives, and the diversity of users and providers in the OECD area. The use and merits of the different types of providers, alone and in combination, are also discussed. Examples of the experience of countries across the OECD area are presented.

Many governments promote the adoption of sustainable farm management practices with the means they have at their disposal. These may be of a regulatory or a voluntary nature, or accompanied by financial incentives. In parallel, farmers and land managers can obtain expertise to improve their resource efficiency, productivity and income, while ensuring compliance with environmental regulations.

Methods used to transfer knowledge to land managers include technical assistance, farm advisory systems and services, training and local action groups. These measures allow them to seize opportunities as well as address concerns and problems related to the use of natural resources. It also helps them to make sound natural resource management decisions. By raising awareness, such measures can change perceptions as to the relevance and performance of a farming practice or innovation, and accelerate the adoption of greener methods.

In the past, these measures would have been considered as agricultural extension and advisory services to individual farmers. They were organised and delivered in different ways, but with the aim of increasing farm productivity and income (Waddington et al., 2010). Indeed, the term *agricultural extension and* advisory *services* was defined by Anderson (2007) as "the entire set of organisations that support and facilitate people engaged in agricultural production to solve problems and to obtain information, skills and technologies to improve their livelihoods".

Within the OECD area, the role of these measures and the way they are delivered and funded have changed. The purpose of agricultural advisory services has been broadened. They are no longer limited to providing technical solutions, but to look more broadly at the institutional environment in which technologies are developed and disseminated, as well as to undertake expert, facilitation, brokering and mediation roles (Christoplos, 2010; Leeuwis and Aarts, 2011; Koutsouris, 2012).

A 2002 review of international literature on information, extension and advisory services and an analysis based on case studies from **Europe**, **North America**, **Australia** and **New Zealand** drew attention to recent developments in advice and information services and the way these are approached, including (Garforth et al., 2002):

- A shift in emphasis from supply-driven extension services – generally organised by public authorities – to services driven by user demand.
- A better understanding of the communication process through which advice is offered and received and information exchanged: this is not a one-way transfer of "messages" but an on-going interaction.
- Recognition that although advice and information services could be funded by the public sector, their delivery can be undertaken by different actors.

Today's understanding of advisory services goes beyond training and sending messages to farmers. It includes assisting farmers to organise and act collectively, addresses processing and marketing issues, and the partnering between a broad range of service providers and institutions. Farmers are considered to be partners in the process of the generation of technology, rather than simply recipients of technology. The range of organisations that provide advisory services has also increased, and includes public sector agencies, non-governmental organisations, and the private sector.

Faure et al. (2012) show that advisory services in all countries are increasingly perceived as an element in an innovation system that leads multiple actors to interact. This perception enlarges the concept of advisory services, which move from being the unique link between farmers and research to one between many diverse actors. In such an advisory system, different advisory service providers fulfil different roles, address different topics and provide different types of advice in different ways.

The wide diversity of approaches used by countries – with each responding to different agro-economic, social and institutional challenges – as well as the evolving nature of these approaches from a linear to more integrated innovation systems were highlighted by the OECD conference on *Improving Agricultural Knowledge and Innovation Systems* (OECD, 2012b). Among the drivers of these changes was the growing disconnection between farmer knowledge and more traditional extension systems on the one hand, and, on the other, a policy agenda that reflected growing concern about the environmental impact of agriculture, the quality of life and employment prospects in rural areas, and the need to support public goods linked to agricultural production (EU SCAR, 2012).

Governments and farmers are now expected to contribute to a growing range of more demanding agri-environmental objectives (e.g. aimed at soil and water management, biodiversity conservation, to increase carbon sequestration, and to reduce greenhouse gas emissions and diffuse pollution), while continuing to produce food, feed and other marketable goods.

Policy tools include regulations, financial incentives (often conditioned on cross-compliance requirements) and voluntary action. It is generally assumed that the uptake of these policies, the level of farmer compliance with land management requirements and environmental outcomes can all be improved through appropriate advisory, training and extension agri-environmental measures. However, the funding and delivery of these measures has changed over time, with the relative decline of public sector dominance and the emergence of a wide range of other actors with differing sets of skills and priorities.

Agricultural advisory services play an important role to support the agricultural sector as an engine of green growth and to enable farmers to meet new challenges, such as the adoption of environmentally sustainable farming practices and improved competitiveness. Studies on the profitability of investment in advisory services have shown relatively high returns. For example, in a meta-analysis of 292 research studies, Alston et al. (2000) found median rates of return of 58% for advisory services investments, 49% for research, and 36% for combined investments in research and advisory services.

Agricultural advisory services are increasingly perceived as a key driver behind innovation processes in agriculture. There is widespread recognition that adoption and uptake of generated knowledge in innovation by farmers (called "organisational innovation") is perhaps the foremost concern in many OECD countries when designing policy approaches to increase agricultural productivity in a sustainable manner (Teagasc, 2013). As challenges such as greenhouse gases (GHG) mitigation and biodiversity loss become increasingly important, as has knowledge generation, and particularly knowledge transfer leading to change in farm practice.

Given these dramatic changes over time in the orientation, organisation and methods of intervention, the issue of how best to provide and finance advisory, training and extension services remains a challenge. The issues include:

- What should be the roles of the public sector, private sector, and civil society?
- How can we ensure that agricultural advisory services are demand-driven and meet the diverse information needs of farmers?
- How can advisory, training and extension services be made efficient and financially sustainable?
- How can we ensure that all farmers have access to agricultural advisory, training and extension services?
- To what extent do such measures contribute to improve the economic viability of a farm, skills and productivity improvement (including on-farm innovation and technology transfer) and to the adoption of environmentally benign farming practices.
- How coherence with other agricultural policy measures can be enhanced?

This report provides a short review of the use of technical assistance, farm advisory systems and services, training and local action groups, illustrated by examples from selected OECD countries. It is a contribution to current OECD work on green growth in the agricultural context, which emphasises the need to accord a higher priority to research, development, innovation, education, extension services and information if productivity is to be increased in a sustainable manner (OECD, 2012c).

Increasing diversity of users and providers

In the context of green growth, the role of advisory, training and extension measures is twofold. First, to improve the capacity of farmers, their advisers and other actors to deliver environmentally appropriate land management over a sustained period, while retaining or improving the economic performance of the farm business (economic role). Second, to encourage and facilitate the adoption of appropriate agri-environmental land management (e.g. compliance with legislation) where this will deliver the desired environmental outcomes (environmental role). In some circumstances, improving the economic viability of a set of practices or holdings may be an environmental goal in its own right; for example, on marginal land at risk of abandonment where sustainable farming is necessary for the provision of environmental and social public goods.

Farmers and land managers are the final consumers of what can be envisaged as a "supply chain" of agri-environmental knowledge originating from many different sources, irrespective of whether they pay directly for the service. The pool of farmer and landowner consumers is diverse, and only a proportion of them will choose to take up these measures. Several studies claim that better educated and trained farm managers are more likely to make successful changes to farm-management practices and become more innovative and flexible (e.g. Charatsary et al., 2011; Labarthe and Laurent, 2009). A particular concern in several countries is the problem of small farmers who may, as a group, be important providers of environmental public goods, but have little or no access to the relevant advisory, training and extension services (see Damianos, 2015). Targeting these measures – by farm size, type of production and farm practice – is clearly important and is discussed below.

It is not possible to estimate current public expenditure on advisory services, training and extension initiatives to support agri-environmental management across OECD countries, but where figures are available the estimates of annual public expenditure are significant, although they remain a very small proportion of total public support for agriculture.[1]

As part of the 2003 CAP reform, the **European Union** introduced the cross-compliance mechanism that linked direct payments to farmers with compliance to basic standards on the environment, food safety, animal and plant health and animal welfare. The reforms also introduced a requirement that farmers maintain land in Good Agricultural and Environmental Conditions (GAEC), with the obligation for member states to set up a Farm Advisory System (FAS). The FAS aims to help farmers to better understand and meet the EU rules for environment, public and animal health, animal welfare and the GAEC. In this respect, as of 2007 national authorities are obliged to offer their farmers advice under FAS, applying certain priority criteria if needed (Council Regulation 73/2009). The rural development policy offers support to farmers that make use of advisory services and to member states to set up new farm advisory services where needed.

Extension systems remain central actors in the delivery of advisory services and training policies in most countries. There is a wide diversity across countries and regions in structure, organisation and governance (including public or private), as well as in the level of centralisation or decentralisation. They are often operating at the sub-national level and include diverse actors: government agencies, education institutions, upstream and downstream industries, non-governmental organisations (NGOs), consultants and farmers' organisations.

An increasing number of services are offered, ranging from technical to financial advice. Government expenditure on extension services in OECD countries, where this occurs, has continued to increase at an annual growth rate of 1% or more with a fairly consistent rate of growth taking place in many countries. The rate slowed down in the **European Union**, **Iceland**, **Korea** and the **United States** in the second part of the 2000s as compared to the first part, but increased in **Australia**, **Chile**, **Israel**, **Japan** and **Mexico** (Figure 2.1) (OECD, 2012a).

Figure 2.1. Government expenditure on extension services

Annual % growth rate, by period, based on USD-PPP 2005

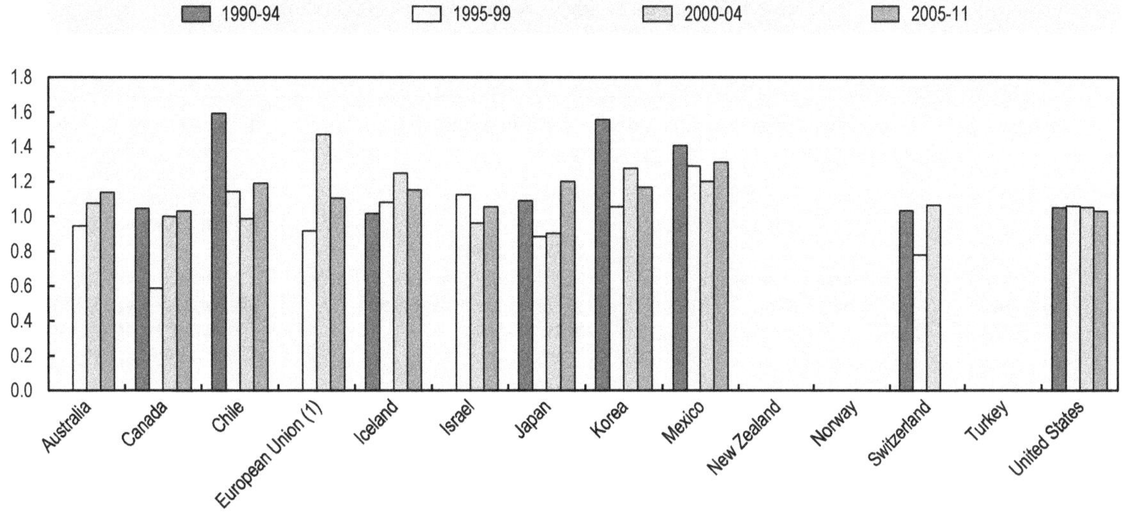

1. EU15 from 1995 to 2003; EU25 from 2004 to 2006; and EU-27 from 2007 to 2011. For the European Union, 2000-03 instead of 2000-04; and 2007-11 instead of 2005-11.

Source. OECD (2012), "Government expenditures on extension services", in *Agricultural Policy Monitoring and Evaluation 2012: OECD Countries,* OECD Publishing, Paris. DOI: http://dx.doi.org/10.1787/agr_pol-2012-en.

Agricultural advisory services, training and extension initiatives are provided and financed by an increasingly diverse range of organisations and a multiplicity of institutional arrangements exist for their financing and provision (Table 2.1). The public and private sectors (farm households, agribusiness enterprises, other profit-oriented firms), and the third sector (non-governmental and non-profit organisations, farmers' organisations, civil society organisations) are involved in different combinations in providing and financing these measures: advice provision involving non-traditional players such as farmers' organisations (financed by the farmers themselves) and non-governmental organisations; contracting out (provided by NGOs or private sector and financed by public sector), privatisation (provided and financed by private sector); and different patterns of public-private partnerships. Within the public sector, institutional structures can differ depending on the degree and the type of decentralisation.

In the **United States**, the extension system was created nearly a century ago to address exclusively rural and agricultural issues, and has evolved over time. Today, extension is achieved through the Co-operative Extension System (CES), a national network of educational institutions in each state which aim to disseminate research-based information on various topics, such as: nutrition, agriculture, small business, youth development and personal finance. Extension focuses on six major areas, one of which is the management of natural resources (teaching landowners and homeowners how to use natural resources wisely and to protect the environment, with educational programmes in water quality, timber management, composting, lawn waste management and recycling).[2] The universities and their local offices are financially supported by the National Institute of Food and Agriculture (NIFA), the federal partner in the CES. Moreover, the interaction among Land Grant Universities and State Agricultural Experiment Stations to provide co-operative extension to agricultural producers (and other actors) is a long-standing institutional arrangement. Greater integration of research and extension has also been a recent emphasis of NIFA programmes.[3]

Some countries (e.g. **Ireland**) retain a public sector advisory service, but recover a proportion of the costs from clients. In the **United Kingdom** (mainly England and Wales), the government's Agricultural Development and Advisory Service (ADAS) was required to charge fees from 1986 and after 11 years of progressive commercialisation, it was privatised in 1997 (Garforth et al., 2002). **New Zealand** and the

Netherlands have also privatised their public sector agricultural advisory services. Outsourcing through various forms of contracting is increasingly common (Rivera and Zijp, 2002).

Table 2.1. Examples of advisory services in OECD countries

	Main institutions	Source of funds	Countries
State-run	Public organisations at regional and national level	Wholly financed from public funds	Belgium, Italy, Greece, Slovenia, Sweden, Germany's southern regions, Spain, Portugal, Luxembourg, Japan, United States, Poland
Public-Private Partnership	Increasingly provided by private consultant firms	Farmers partly or wholly pay for services; centralised and decentralised	Canada, Ireland, Czech Republic, Finland, Slovak Republic, Hungary, Australia, Chile, United States
Farmers' organisations	Farmers' organisations	Membership fees and payments by farmers	Austria, France,[1] Denmark, North-West regions of Germany, New Zealand (e.g. DairyNZ), Norway, Canada, United States
Commercial	Commercial firms or private individuals	Payment through project implementation or grants	England, Netherlands, north-east regions of Germany, New Zealand, United States

1. Advisory services are provided primarily by the Chambers of agriculture (see Box 2.1).

Source: OECD (2012), "Advisory services in OECD countries", in *Agricultural Policy Monitoring and Evaluation 2012: OECD Countries*, OECD Publishing. DOI: http://dx.doi.org/10.1787/agr_pol-2012-en (Table 2.5) (Adapted from Laurent, Cerf and Labarthe, 2006, using response to an OECD questionnaire (www.oecd.org/agriculture/policies/innovation).

The **European Union uses** diverse approaches to implement FAS regulations (ADE, 2009). FAS services, that is services related to GAECs, are implemented by public, private or both types of actor, sometimes free of charge or (partially) financed by farmers. Services are delivered through individual and/or group advice methods, training, or through ICTs.

There are 150 independent Chambers of Agriculture in Europe that provide extension and advisory services for more than 5 million farmers in 14 countries (**Austria**, Croatia, **Czech Republic**, **Estonia**, **Flanders**, **France**, **Germany**, **Hungary**, Latvia, **Lithuania**, **Luxembourg**, **Poland**, **Slovakia** and **Slovenia**).[4] The case of chambers of agriculture in **France** is briefly presented in Box 2.1.

Box 2.1. Chambers of Agriculture in France

Chambers of agriculture in France are professionally managed public institutions. They fulfil a dual mission: they represent agricultural and rural communities to public authorities and play an essential service role for farmers. Since their creation in 1920, their mission, objectives and financing have evolved considerably. Currently, there are 110 such establishments – at the *Départmental*, regional and national levels – employing a total of 7 800 people.

Their main functions include the following: provision of individual business advice for farmers (commercial strategy, organisation, investment in equipment), agronomic and environmental advice, territorial and local development, quality of products, and the monitoring of intangible resources and data bases. Some of these services are mandatory missions associated with the delegation of service from the Ministry of Agriculture, Agri-food and Forestry (MAAF). They also have a training centre and an agricultural private college.

The Chambers are chaired by a president, and a board of farmers' representatives nominates the director. They combine different sources of funding: a local tax on "non-built" land (on average 50% of the funding of chambers), public funding from the MAAF (about 17% of total funding), contracts with local authorities (regions, departments), and the purchase of services by farmers.

The decentralisation of public extension services has led to the emergence of private actors including consultant agronomists and land agents employed by farmers, but there are many other types of providers (Laurent and Labarthe, 2011). Environmental and agricultural non-governmental organisations (NGOs), farmer co-operatives and informal self-help peer groups are well suited as knowledge brokers.[5] In the commercial sector, suppliers of inputs such as feed, fertilisers and machinery are increasingly strengthening customer loyalty by giving advice to farmers. Alternative sources of advice such as these are relevant in countries where extension services and co-operation are relatively weak, such as **Italy** and Latvia (EU SCAR, 2012).

The trend away from the delivery of advice by government departments and agencies to various combinations of private and public funding of services delivered by private sector organisations is perhaps the most discernible change concerning the provision of these measures over the last two decades.[6] This may originate from a drive to reduce public expenditure and to transfer costs from the state to the final beneficiaries in order to better align the supply of these services with the perceived needs of users and to increase their effectiveness. In the *fee-for-service system* that characterises most private provisions, the type of information that is of priority to farmers could be more clearly revealed.

The privatisation trend has resulted in a huge diversification of the service provision itself with different types of private advisory firms and various patterns of public-private partnerships (Rivera and Zijp, 2002). The disengagement of governments and the emergence of the private sector raise the question of the sustainability of advisory services and, therefore, of their funding. It has also led to an interesting debate as some unanticipated effects of state withdrawal have influenced the effectiveness of AKIS in many countries (Kidd et al. 2000; Laurent, Cerf and Labarthe, 2006).[7] These concern:

- Access to advisory and extension services for some categories of farmers (e.g. less commercial farmers) and their ability to finance advice as the price of the service will tend to be market-determined.
- The adequacy of knowledge flows between the various stakeholders of AKIS, including research and extension.
- Existence of externalities (e.g. related to soil conservation) that most likely imply social inefficiencies (Faure, et al. 2012).[8]

An important consequence of the state's withdrawal from the funding and managing of agricultural advisory services is that the content of the knowledge produced and made available in "back-office" activities (e.g. re-organisations concerning R&D, training advisors, and production, collection and storage of technical knowledge) have changed. Labarthe and Laurent (2013) argue that the disengagement of the state has resulted in major transformations of back-office activities including, *inter alia*, the increasing importance of upstream industries that finance a large number of trials, less feedback between suppliers and users agricultural extension services and commercialisation of advisory services by applied research institutes in a highly segmented way to different clients, including upstream industry.

This evolution may be severely detrimental to the interests of some types of farms, in particular small-scale farms or those with innovative production systems. Research in **Finland, France, Germany**, **Italy**, **Latvia**, the **Netherlands** and **Switzerland** found that specific groups of farmers lack access to support services (Knickel et al., 2009). Smaller or extensively managed farms and those below certain output thresholds may not qualify for government extension programmes that are designed largely for more intensive systems, or find it difficult to access these because of costs or gaps in territorial coverage (EU SCAR, 2012). For example, uncompetitive farms in urban and agriculturally marginal areas in **Italy** tend not to use extension services, although these farms may play a significant role in producing environmental public goods (De Rosa et al., 2012).

Labarthe and Laurent (2013) advocate that if the current trend continues, this knowledge could become increasingly less relevant to small farms, as a result of the decreasing demand from those farms and the increasing dominance of large producers.[9] Investment in the "back-office" is an important issue for agricultural innovation policy and plays a crucial role in assuring that knowledge produced is reliable and accessible to all.

Targeting these measures by types of users could address the concern of access by smaller and disadvantaged farmers. For instance, smaller-scale and poorer farmers in **Chile** are served by public extension or by contract extensions through which larger shares of public funding are received (e.g. an association of smaller farmers receives a larger matching allocation to hire extension staff) (Rivera and Alex, 2004). **Portugal**, for the implementation of the EU FAS regulation, has chosen to target social groups, such as women or young farmers.

The privatisation of advisory services also implies that the state should both develop new functions to regulate relationships between stakeholders and ensure that public interests are considered (Labarthe, 2005; Klerkx et al., 2006; Rivera and Alex, 2004). Policy-makers can create enabling conditions for the emergence of advisory services that are financed and managed by the private or the third sector (farmers' organisations, agri-business enterprises, etc.).

Some authors consider that the state has a role to play above all in the most disadvantaged areas and for the poorest farmers (e.g. Anderson and Feder, 2004 and 2007; Kidd et al., 2000). As Garforth et al., 2002 indicate the growing acceptance that public funding of support measures should be focused on the provision of public goods from agriculture. In practice, the relative importance of public and private funding varies considerably, reflecting government policies, type of provider, and the purpose of the measure.

As emphasised by Rivera and Zijp (2002), the evolution towards a privatised advisory system is not straightforward and requires precise clarification of the role of institutions, economic opportunities to be able to finance advisory services, advisory service providers with adequate capacities, and that farmers are able to formulate clear demands. In general, advice provided by the private sector, including farmer organisations, in terms of content, accountability, quality, investment in R&D, and cost merit further analysis (Faure et al. 2012).

Notes

1. For example, approximate annual expenditure of around EUR 145 million on the FAS in the **European Union**, of AUD 55 million on Landcare in **Australia**, and of USD 797 million on the Conservation Technical Assistance Programme in the **United States**.
2. http://www.csrees.usda.gov/qlinks/extension.html.
3. See, for example, http://www.csrees.usda.gov/funding/integrated/integrated_synopsis.html.
4. EU Chambers of agriculture (2011) quoted in EU SCAR (2012).
5. The literature on knowledge brokers provides many illustrations of various initiatives to recreate networks between farmers, advisory and research organisations in order to support innovation (Klerk and Leeuwis, 2008).
6. In the **European Union**, from the end of the 1980s, a general tendency towards state withdrawal from the funding and management of extension services started to appear on a national scale. This was initiated in the **United Kingdom** and the **Netherlands**.
7. Leeuwis (2000) and Labarthe (2005a) in the **Netherlands** report severe difficulties engendered by privatisation with regard to the dissemination of complex innovations involving environmental concerns and production systems; a fragmentation of advisory service themes; a priority to linear transfer of innovation; a reduction of information

exchange between farmers; and a selection of financially solvent farmers by advisory service providers which could lead to an exclusion process.

8. Several authors mention the drawbacks of privatisation without going so far as to reject this option. This is, for instance, the case of Kidd et al. (2000), whose article is based on different case studies throughout the world. See also Marsh and Pannell (2000) in **Australia**, and Klerkx, et al. (2006) in the **Netherlands**.

9. Labarthe and Laurent (2013) found that in certain European farmers' co-operatives for instance, it was observed that they lead to a segmentation of both front- and back-office services. These cooperatives offer farmers "package" solutions that combine some technological options (in terms of seeds varieties and chemicals) and some level of interaction with farmers (field visits with advisors which vary in number and purpose). But these packages rely on different levels of investment in back-office activities. For larger farms, they offer highly flexible solutions that are characterised by intensive front-office interaction, as well as by intensive R&D experiments. For smaller farms, they tend to offer more standardised solutions, with less interaction. This often leads to lower yields as the solutions preferred are rarely based on recently updated knowledge of back-office activities.

Bibliography

ADE (Aide à la décision économique) (2009), *Evaluation of the Implementation of the Farm Advisory System*, ADE in collaboration with ADAS, Agrotec and Evaluators EU, Belgium. http://ec.europa.eu/agriculture/eval/reports/fas/index_en.htm.

Alston, J., M. Andersen, J. James and P. Pardey (2011), "The Economic Returns to U.S. Public Agricultural Research", *American Journal of Agricultural Economics*, Vol. 93, No. 5.

Alston, J., C. Chan-Kang et al. (2000), *A Meta Analysis of Rates of Return to Agricultural R&D: Ex Pede Herculem?,* Research Report 113. International Food Policy Research Institute, Washington, DC.

Anderson, J. (2007), *Agricultural Advisory Services*, Background Paper for the World Development Report 2008, http://siteresources.worldbank.org/INTWDR2008/Resources/2795087-1191427986785/Anderson_AdvisoryServices.pdf.

Anderson J. and G. Feder (2007), "Agricultural extension", In R.E. Evenson and P. Pingali (eds.), *Handbook of agricultural economics*, Vol. 3, Elsevier Science, Amsterdam.

Anderson, J. and G. Feder (2004), "Agricultural Extension: Good Intentions and Hard Realities", *The World Bank Research Observer*, Vol. 19, No. 1.

Charatsary, C., A. Papadaki-Klaudianou and A. Michailidis (2011), "Farmers as consumers of agricultural education services: willingness to pay and spend time", *Journal of Agricultural Education and Extension*, Vol. 3, No. 3.

Christoplos, I. (2010), *Mobilizing the Potential of Rural and Agricultural Extension*, Food and Agriculture Organization Publications, Rome.

Damianos, D. (2015), "An Analysis of the Performance of Farm Advisory Services as Related to Fostering Green Growth in Agriculture – The Case of Greece", Consultant Report, OECD internal document.

European Union (EU) SCAR (2012), *Agricultural Knowledge and Innovation Systems in Transition – a reflection paper*, Brussels, http://ec.europa.eu/research/bioeconomy/pdf/ki3211999enc_002.pdf.

EU SCAR (2013), *Agricultural Knowledge and Innovation Systems Towards 2020 – An Orientation Paper on Linking Innovation and Research*, Brussels, http://ec.europa.eu/research/bioeconomy/pdf/agricultural-knowledge-innovation-systems-towards-2020_en.pdf.

Faure, G., Y. Desjeux and P. Gasselin (2012), "New Challenges in Agricultural Advisory Services from a Research Perspective: A Literature Review, Synthesis and Research Agenda", *Journal of Agricultural Education and Extension*, Vol. 18, No. 5.

Garforth C, B. Angell, J. Archer and K. Green (2002), *Improving access to advice for land managers: a literature review of recent developments in extension and advisory services*, DEFRA Research Project KT0110, Department for Environment, Food and Rural Affairs. London.

Kidd, A., J. Lamers et al. (2000) "Privatising agricultural extension: caveat emptor - a sectoral analysis with some thoughts on accountability, sustainability and evaluation", *Journal of Rural Studies*, Vol. 16, No. 1.

Klerkx, L. and C. Leeuwis (2008), "Matching demand and supply in the agricultural knowledge infrastructure: experiences with innovation intermediaries", *Food Policy*, Vol. 33, No. 3.

Klerkx, L., K. de Grip and C. Leeuwis (2006), "Hands off but Strings Attached: The Contradictions of Policy-induced Demand-driven Agricultural Extension", *Agriculture and Human Values*, Vol. 23, No.2.

Knickel K, T. Tisenkopfs and S. Peter (eds.) (2009), *IN-SIGHT: Strengthening Innovation Processes for Growth and Development. Innovation processes in agriculture and rural development: results of a cross-national analysis of the situation in seven countries, research gaps and recommendations.* EU Sixth Framework Programme: Priority 8.1 policy-oriented research, scientific support to policies – SSP: 44510. www.insightproject.net/files/IN-SIGHT_final_report.pdf.

Koutsouris, A. (2012), ''Facilitation and Brokerage: New Roles for Extension'', *Journal of Extension Systems*, Vol. 28, No. 1.

Labarthe, P. (2005), "Trajectoire d'innovation des services et inertie institutionnelle: dynamique du conseil dans trois agricultures européennes", *Géographie, Économie, Société*, Vol. 73, No. 3.

Labarthe, P. and C. Laurent (2013), "Privatization of agricultural extension services in the EU: Towards a lack of adequate knowledge for small-scale farms?" *Food Policy*, Vol. 38.

Laurent, C. and P. Labarthe (2011), "Économie des services et politiques publiques de conseil agricole", *Cahiers Agricoles*, Vol. 20, No. 8 5, September-October.

Labarthe, P. and C. Laurent (2009), *Transformations of agricultural extension services in the EU: towards a lack of adequate knowledge for small-scale farms.* Paper presented at the 111 EAAE-IAAE seminar "Small farms: decline or persistence", University of Kent, 26-27 June. http://ageconsearch.umn.edu/bitstream/52859/2/103.pdf.

Laurent, C., M. Cerf and P. Labarthe (2006), "Agricultural Extension Services and Market Regulation: Learning from a Comparison of Six EU Countries", *Journal of Agricultural Education and Extension*, Vol. 12, No. 1.

Leeuwis, C. (2000), ''Learning to be Sustainable, Does the Dutch Agrarian Knowledge Market Fail?'' *Journal of Agricultural Extension and Education*, Vol. 7, No. 2.

Leeuwis, C. and N. Aarts (2011), ''Rethinking Communication in Innovation Processes: Creating Space for Change in Complex Systems'', *The Journal of Agricultural Education and Extension*, Vol. 17, No. 1.

Marsh S. and D. Pannell (2000), "Agricultural extension policy in Australia: the good, the bad and the misguided", *Australian Journal of Agricultural and Resource Economics*, Vol. 44, No. 4.

OECD (2012a), *Agricultural Policy Monitoring and Evaluation 2012: OECD Countries*, OECD Publishing, Paris.
DOI: http://dx.doi.org/10.1787/agr_pol-2012-en.

OECD (2012b), *Improving Agricultural Knowledge and Innovation Systems - OECD Conference Proceedings*, OECD Publishing, Paris.
DOI: http://dx.doi.org/10.1787/9789264167445-en.

OECD (2012c), *Food and Agriculture*, OECD Green Growth Studies, OECD Publishing, Paris.
DOI: http://dx.doi.org/10.1787/9789264107250-en .

Rivera, W.M., M.K. Qamar and H.K. Mwandemere (2005), *Enhancing coordination among AKIS/RD actors: an analytical and comparative review of country studies on agricultural knowledge and information systems for rural development (AKIS/RD)*, FAO, Rome.

Rivera W. and G. Alex (2004), "Extension system reform and the challenges ahead", *Journal of Agricultural Education and Extension*, Vol. 10, No. 1.

Rivera, W. and W. Zijp (2002), *Contracting for Agricultural Extension. International Case Studies and Emerging Practices,* CABI Publishing, Cambridge, MA.

Rivera, W. (2000), "Confronting global market: public sector agricultural extension reconsidered", *Journal of Extension Systems*, Vol. 16.

De Rosa, M, L. Bartoli and S. Chiappini (2012), *The adoption of agricultural extension policies in the Italian farms*, Paper prepared for the 126th EAAE Seminar: *New challenges for EU agricultural sector and rural areas. Which role for public policy?* Capri, Italy, 27-29 June. http://ageconsearch.umn.edu/bitstream/126155/2/DeRosa%20Bartoli,Chiappini.pdf.

Swanson, B.E. (2006), "The changing role of agricultural extension in a global economy", *Journal of International Agricultural and Extension Education*, Vol. 13, No. 3.

Teagasc (2013), *Knowledge Transfer Conference 2013: Future of Farm Advisory Services, Delivering Innovative Systems*, Dublin, www.teagasc.ie/events/2013/20130612.asp.

Waddington, H., B. Snilstveit et al. (2010), *The Impact of Agricultural Extension Services*, 3ie Synthetic Reviews – SR00 9 Protocol, International Initiative for Impact Evaluation, www.3ieimpact.org/media/filer/2012/05/07/009%20Protocol.pdf.

Chapter 3

Types of advisory services, training and extension initiatives

This chapter examines different types of advisory, training and extension measures, focusing on technical advice, agri-environmental knowledge transfer between researchers and farmers/advisers, training in agri-environmental management, peer group and co-operative initiatives and measures based on information and communications technology.

Measures to promote advisory services, training and extension initiatives in the OECD area are broad in their coverage of environmental objectives and types of farm business management. OECD countries share the premise that a potential barrier to uptake for many farmers and other actors is their lack of familiarity with both the concept and the practicalities of environmental management.[1]

The need to overcome this should not be underestimated, not just for farmers but for other providers whose expertise lies in agricultural production methods rather than environmental land management. When these measures are introduced, there is often a need to first "train the trainers and advisors" and thereafter ensure they have access to up-to-date sources of knowledge.

The great majority of advisory, training and extension measures are voluntary; participation or engagement depends on the farmer, land manager, advisor or other actor *choosing* to use the opportunities available.[2] These measures must therefore have some level of appeal, typically offering one or more elements which can be seen as useful by the target group.

Clearly, the value of participation for individual farmers will depend on many factors, including their existing knowledge, perceptions and attitudes, as well as the agricultural, economic and environmental aspects of their business (Chapter 4). There are several types of advisory, training and extension measures which may be offered as stand-alone options or in different combinations or sequences for particular purposes, context or target groups of "consumers". For this report, they have been considered in the following four groups, although in practice the distinctions may not be so clear-cut.

- Technical advice and support.
- Training in agri-environmental management.
- Peer group and co-operative initiatives.
- Measures based on information and communications technology (ICT).

Technical advice and support

Providing effective technical advice and support to introduce or maintain agri-environmental management is not simply a matter of ensuring that farmers are given detailed technical information on the type of land management required and any changes this may involve. It is a more complex process which can only begin with the farmer's acceptance of the need to adopt the relevant form of agri-environmental management. The technical support process can be broken down into a series of complex and overlapping pathways, including connections between education, research and extension, and that knowledge and impacts are able to move in multiple directions (e.g. from farmers back up to researchers through extension services, instead of a unidirectional flow from extension to farmers) (Chapter 4).

Five key stages in the provision of technical advice for agri-environmental management can be identified: i) making farmers aware of the need for and benefits of agri-environmental action; ii) helping individual farmers assess relevance to and impact on their farm business; iii) supporting farmers in the decision-making and application process, where relevant; iv) helping farmers, their workers and advisers to understand and implement changes; and v) provide technical "after care" and enable farmers to assess environmental outcomes. Technical support and advice for the first stages may be generic and delivered through printed material, codes of practice, group workshops, and via Internet sites; one-to-one meetings between advisers and farmers are nevertheless usually more effective and become more important in the latter stages when farmers need to make detailed decisions about their own land.

Box 3.1. Farm Advisory Services (FAS) in EU member states

As a condition to receive direct payments under the Common Agricultural Policy (CAP), EU farmers are required to comply with specific standards of good agricultural and environmental management, and with certain regulations on the environment, animal health and welfare. Since 2007, the 27 member states have been legally obliged to set up a national Farm Advisory System (FAS) with the broad objective to help farmers meet cross-compliance standards. In around half of the Member states, the FAS is set up as a specific service, complementing existing extension services; in the others, the FAS has been integrated with existing services. Agencies delivering the service to farmers are selected by calls for tender in 14 member states, and by designating private or public providers (five Member states in each case). Most countries set the minimum threshold for advisors' qualifications at university level (BSc or MSc).

A farmers first contact with the FAS is usually via telephone help lines, but on-farm one-to-one advice is the approach most widely adopted, complemented by on-farm small group discussions. Computer-based information tools and checklists are used in several countries. One-to-one on-farm advice is provided free of charge in some countries, others require the farmer contribute to the costs (this varies from 20 to 100%). Although raising farmers' awareness of cross-compliance standards is the main purpose of the FAS, Member states can choose to include advice on other issues. Around half do so, offering advice on broader issues such as the competitiveness of the holdings, the environmental impact of farming practices, and support for implementation of rural development measures such as agri-environmental contracts.

To date, there has not been much prioritising of groups of farmers as key recipients of advice, except the initial obligation for Member states to give priority to farmers receiving more than EUR 15 000 in direct payments per annum. The main beneficiaries of the FAS have been large farms, already familiar with the existing advisory services. Authorities in some Member states have stated they have had problems in reaching small farms. In the EU as a whole, the number of farmers receiving FAS advice has been limited (around 5% of farmers receiving direct payments were given one-to-one advice in 2008).

Member states can also receive part payment under the rural development component of the CAP for expenditures arising from providing financial support for farmers to use advisory services. It is estimated that twenty EU member states have supported more than a million farmers in this way during 2007-13, with a budget of EUR 871 million (EC, 2010).

For the 2014-20 period, it is foreseen that the scope of the FAS will be broadened to, *inter alia*, actions related to agricultural practices beneficial for the climate, innovation and to the promotion of entrepreneurship.

Source: http://ec.europa.eu/agriculture/direct-support/cross-compliance/farm-advisory-system/index_en.htm; http://eur-lex.europa.eu/LexUriServ/LexUriServ.do?uri=OJ:L:2013:347:0549:0607:en:PDF.

In the **EU** Member states, for example, several methods are used but there is little targeting of individual farmers in delivering the EU wide cross-compliance Farm Advisory Services (FAS) (Box 3.1).[3] Some countries have chosen not to specify any target population (e.g. **France**, **Poland** and many **German** Länder), while others have focused their strategy on the bigger farms (e.g. the **Netherlands** and **Denmark**) or on the contrary on smaller farms (e.g. Romania). Certain countries have targeted some social groups, such as women or young farmers (e.g. Bulgaria and **Portugal**); in the **United Kingdom**, the FAS pattern has included a zoning rationale (linked with Nitrogen Vulnerable Zones or areas targeted for phytosanitary emergency plans).

In contrast, some agri-environment advice and support is aimed at specific, sometimes localised, land management and environmental goals, often in tandem with targeted incentive measures. Greater complexity and focus may create the need for an interactive process, especially where site-specific questions are likely to arise. Two examples relating to the management of particular bird species and one example relating to water protection and environmental conservation are presented in Box 3.2.

Where a farmer needs advice on how to integrate environmental management with the rest of his business, targeted advice may be an important element in the decision making process. In this case, skilled and objective advice may be necessary to identify synergies and resolve potential conflicts between the policy objective of maximising environmental benefits and the business needs of the farm. Such advice is often, but not always, publicly funded.

Box 3.2. Technical advice targeted at conservation of farmland birds in Scotland (UK), Belgium and agri-environmental advisory in South West Finland

In Scotland (**United Kingdom**), pastureland provides important breeding sites for wading birds such as curlew, lapwing and redshank. Within the government agri-environmental programme there are targeted options for the management of wetland and grassland that allow for drawing up dedicated land management plans for particular sites, but farmers require specialist help to do this. The Strathspey Wetlands and Waders Initiative was launched to address the problem of a sharply declining local wader population in the semi-natural floodplain of the Spey, where numbers of breeding birds had fallen by 42% between 2000 and 2010. The aim was to foster collaboration between conservationists, agriculturalists and land managers to produce high quality land management plans, taking a strategic approach to the development of habitats at a landscape scale, and to encourage uptake of agri-environment management payments. Landowners and tenant farmers are offered very specific on-the-ground advice about breeding bird habitats, for example required grass height for feeding, the proportion of the pasture where rush (Juncus) is allowed to grow and the number of shallow pools needed. There are also other forms of capacity building such as training, research, and networking. Technical advice is provided by a combination of actors, including the Scottish Agricultural College, Cairngorm National Park Authority, independent agricultural agents and the Royal Society for the Protection of Birds (RSPB), an environmental NGO. Results collated in February 2012 showed 2 250 hectares of land registered within the Initiative but, despite this positive uptake and improved management of the wetlands, the wader population is still in decline and the RSPB is concerned about reduced availability of future agri-environmental funding until the current CAP reform process is completed.

The Flemish region in **Belgium** offers another example of focused information and advice to farmers aimed at addressing declining farmland bird populations. In contrast to the UK example, this is a bottom-up approach established by an informal cooperation between regional consultants, farmers, millers, bakeries and bakery schools. The project, *BakkerBrood* sought to increase cereal producers' awareness about the benefits of adopting certain land management practices to ensure the provision of winter feed for farmland birds. Technical support included training, field demonstrations and online guidance documents for farmers. In common with the Strathspey Wetlands and Waders Initiative, *BakkerBrood* had a positive reaction from land managers and a good uptake, but in this case the outcome for farmland birds is more encouraging, with increases in populations of skylark and corn bunting.

A key objective of the TEHO Plus-project, which operates in the south-western parts of **Finland**, is the development of environmental counselling services to raise farmer awareness and develop new ways of targeting environmental measures, in order to increase efficiency, particularly in the areas of nutrient balance, varied crop rotation, biodiversity, energy consumption and soil texture. These have been provided to farmers free of charge. During 2008-13, over 200 counselling visits to 175 farms took place. The project, which is a joint project between the Union of Agricultural Producers and the Center for Economic Development, Transport and the Environment in south-western Finland, was funded by the Ministry of the Environment and the Ministry of Agriculture. Evaluation of the project found that the counselling services: i) led to a change in farmers' attitudes; ii) increased the uptake of the agri-environmental support scheme in the region (e.g. buffer zones); and iii) resulted in a positive net social benefit (i.e. the benefits to society coming from the counselling services and the environmental benefit – in this case buffer zones – are higher than the costs of the services). It was also noted that high quality counselling, combined with information and targeted environmental measures designed for each individual farm is the most effective way of achieving a positive environmental impact. This also contributes to changing attitudes, especially in the case where farm specific measures benefit both the environment and the farm's economy. New ideas were seen to be better accepted when coming from an experienced counsellor, or even better a colleague or a neighbour. An additional advantage of the project is the cooperation between the producer union and the regional environmental authorities, traditionally two opposing partners.

For example, the carbon accounting tool (CALM) in **England** is provided free by a landowners' association and a consultancy firm (Box 3.3). In a public-private collaboration in south-east **Norway**, the Ministry of Food and Agriculture, the Norwegian Farmers' Union and agricultural advisors set up a catchment-based pilot project to address the pollution in Lake Vansjo (South East Norway) caused by phosphorus leaching from farms in the catchment (Øgaard, 2011).

This financial support for the adoption of land management restrictions and the key role of the agricultural advisors was to provide "one-on-one" advice to farmers in the preliminary stages of environmental planning and drawing up of contracts. Advisors also educated farmers about the importance and impacts of the mitigation practices they were adopting. This advice was considered to play an essential role in the high level of farmer participation.

The farmer implementing a government-funded conservation programme may not distinguish between the related but differing needs for administrative and technical support. However, this distinction is important from the point of view of effective resource allocation because the two functions often require different skills and knowledge. In an effort to make the most effective use of staff resources, the Natural Resources Conservation Service (NRCS) in the **United States** initiated in 2009 the Conservation Delivery

Streamlining Initiative. The aim is for technical field staff to spend as much as 75% of their time in the field with customers and to eliminate, automate or reassign 80% of the administrative assistance tasks they currently spent time on (Stubbs, 2010).

Box 3.3. CALM: An interactive carbon management tool in the United Kingdom

Carbon Accounting for Land Managers (CALM) is a free online tool that farmers and land managers can use to calculate net greenhouse gas (GHG) emissions from farms. CALM was developed in **England** (UK) by a landowners' association (the Country Land and Business Association) in partnership with the private land agency Savills. The tool provides farmers with an opportunity to engage with technical information and informs them about the greenhouse emissions of their business and the potential for carbon sequestration so that they can react accordingly to reduce and offset emissions. An important element of this service is that it does encourage farmer feedback on performance.

An example of the CALM calculator in use is the Alwinton sheep farm within the Northumberland National Park in northern England. The CALM calculator showed that aggregate GHG emissions were 3.2 CO_2e/ha. It demonstrated how planting an extra 13 hectares of woodland in addition to the 80 hectares already present would offset the farm's estimated carbon emissions through sequestration and achieve a "carbon neutral" farm.

One criticism of the CALM calculator is that it is confined to farm level operations and so could be made more accurate by accounting for emissions from production of inputs upstream of the farm, such as livestock feed and fertiliser (Bright et al., 2008).

Source: http://www.calm.cla.org.uk/.

Agri-environmental knowledge transfer between researchers and farmers/advisers

It is vital that providers of advisory services, training and extension initiatives for land management are informed and kept up-to-date with the best available environmental management techniques. This requires a process of knowledge transfer between researchers on the one hand, and advisers and practitioners (farmers) on the other. As discussed in Chapter 4, agricultural advisory services, training and extension measures are part of the wider AKIS systems in which knowledge and innovations are generated, disseminated, and utilised in the agricultural sector.[4]

However, according to recent research in the **United Kingdom**, research funding does not typically cover dissemination of the results to stakeholders or target groups, and potentially valuable research is not taken further than the publication of results in academic journals. Research institutions, programmes and projects often could make better use of advisors in knowledge exchange, and professional organisations, universities, colleges and other training institutes need to reconsider the type of skills that field-based advisors require (RELU, 2011a).

Knowledge transfer between researchers and farmers requires some form of mediation to avoid misinterpretation and misunderstanding. One way is through on-site-research and end-user involvement in research dissemination, removing the potential inefficiency of researchers developing impractical management methodologies and farmers misinterpreting technical guidance (Garfoth, 1998). The successful examples of two-way information flows shown in Box 3.4 involve collaboration between many different actors, including two examples from the **United States** where the aim is to use the results of on-farm research and innovation to inform best practice on other farms.[5]

Demonstration farms can be an effective means for knowledge transfer and to promote changes in how farmers operate (LEAF, 2009). Evidence from the **United Kingdom** shows that such schemes – which assist learning, confidence building and motivation – are valued by farmers and more likely to bring about long-term benefits than those which provide information only (Defra, 2002). Showing someone as an exemplar for good environmental behaviour can create a ripple effect, whereby farmers want to achieve the same success (Slee et al., 2006).

Box 3.4. Improving information flows between researchers and farmers

Climate Farmers in the **Netherlands** was developed by two young farmers' organisations keen to bridge the knowledge gap between researchers and farmers. Researchers had developed numerous techniques to reduce GHG emissions, but uptake among farmers was poor. There were difficulties implementing the proposed techniques on farms, and some were not cost-effective, and the project sought to adapt the research results in an accessible manner for farmers. It also collates examples from dairy and arable farms of best farming practice to reduce GHG emissions. The information is presented in a short online booklet outlining the practices and providing case study examples in each instance (Climate Farmers, 2012). This online resource is available to all farmers but it specifically targets young European farmers.

In **Israel** the government and farmers' associations collaborated in a recent research initiative to create a database of regional economic and agronomic statistics, for example farm size and farm structure. The principle aim of this research by the Institute of Farm Income Research was to strengthen economic capacity of Israeli farmers. The participation of farmers' associations in this instance played a key role by ensuring farmer cooperation in collecting the data.

In the **United States** the Iowa Soybean Association, runs an *On-Farm Network* that assists farmers in organizing and conducting on-farm research about nutrient use, to document changes in the efficiency of nitrogen use on crops. The goal is reducing nitrogen applications to achieve positive environmental effects and lower input costs. The beneficial management practices resulting from this on-farm research are then presented to other members of the Association.

Since 2004, *Conservation Innovation Grants* (CIG) have been awarded in the United States to stimulate innovative approaches to environmental enhancement and protection on agricultural land. The results are intended to provide a return on federal investment, by incorporating the findings into the NRCS consortium of technical tools available. A recent report to Congress notes that technical assistance has been based historically on science-based principles and asks how the technology transferred from individual CIG projects has been incorporated into the national technical assistance effort, and whether this helped or hindered producer application of new technology through federal programmes (Stubbs, 2010).

Conservation Evidence is a free online resource (journal) developed by scientists to collate and summarise the scientific evidence supporting the use of specific agri-environmental management techniques on farmland (www.ConservationEvidence.com). For example, the database covers issues such as creating uncultivated margins around intensive arable or pasture fields, converting arable land to permanent grassland, connecting areas of semi-natural habitat and providing wildlife refuges during harvest or mowing.

Training in agri-environmental management

Training in specific agricultural and business skills has long been an important element of agricultural extension services, and now has a new role in environmental land management. Training often involves on-site interactions with land managers, field days, forums, group workshops, and instructive pamphlets. Technology is increasingly playing a more significant role in training, for example training online and via mobile phones.

In the **European Union**, training has been used by some member countries as a measure to improve the competiveness of the agricultural sector and sustainable management of natural resources (Dwyer et al., 2012). For example, vocational training (under Axis 1: improve the competitiveness of the agriculture and forestry sectors of Pillar 2 of the Common Agricultural Policy) includes training in the more efficient use of fertilisers and which is provided in **France**, **Austria**, and **Italy**; this has been made a prerequisite to receive other kinds of support (e.g. farm modernisation support and support to young farmers) in order to ensure adequate farmer uptake. Training and information relating to energy efficiency, opportunities for generating renewable energy and developments in relation to environmental technology are also provided (e.g. in **Belgium**, the **Netherlands**, **Sweden** and the **United Kingdom**).

Recent research in the **United Kingdom** has shown that training programmes help to improve the outcome of voluntary agri-environment schemes by changing the attitudes of participating farmers. Rather than focusing on payments, they are more engaged in the conservation objectives; they have a better understanding of the reasons that underlie the actions they are asked to take, as well as how to achieve them; and training programmes ensure that farmers have all the information and skills they need to make their interventions effective.

In **Canada**, an objective of Environmental Farm Plan programme is to engage producers and improve their awareness of agri-environmental issues and best management practices to encourage adoption beyond those that are financially incentivised. Research in some countries suggests that it is important that training is focused on practical issues and delivered by professionals with farming experience who can gain the respect of farmers (RELU, 2011b).

Training is not just for land managers. The quality of advisory services greatly depends on the skills of advisors, such as their capacity and ability to deliver advice that is appropriate, relevant and easily understandable by the beneficiary.

Attracting well-qualified advisors with diverse and flexible skills into the agriculture sector is a challenge and there is a need to train agri-environment advisors and agronomists too. As advice is increasingly moving from supply to demand driven, advisors are required to develop new skills in line with farmers' changing requirements. On the one hand, there is a need to extend the skills of advisors to deal with a broad spectrum of topics, while on the other hand specialised training needs to be provided to obtain highly qualified advisors to deal with specific topics.

Advisors need access to the best available scientific and technical knowledge. This is done in "back-office" activities, when advisors take courses, update external knowledge bases, build databases, and review scientific experiments and scientific and technical literature, etc. Both the mobilisation and the production of this knowledge may involve various institutional co-operation patterns, as described in several approaches (Klerxk and Jansen, 2010; Labarthe, 2005).

In a study of innovation in agriculture and rural development, Knickel et al. (2009) argue there is often a gap between the need for change and a farmers' willingness to adjust on the one hand, and a lack of capacity to support change within the agencies responsible for innovation and advisory services on the other. They emphasise that institutions, administrations and extension services can become barriers to progress if these fail to acknowledge that the needs of farmers and of society have changed.

Focusing on an analysis of the advisory profession, Remy et al. (2006) illustrate the diversity and range of new knowledge and skills required of agricultural advisors in France to manage concrete issues (production, management, administration, etc.) and the interpersonal dimensions of advisory services.

Garforth (2011) points out that staff brought up in the "technology transfer" tradition may need re-orienting towards a more participatory, interactive approach. In a study of young farmer training in **Greece**, Labarthe and Laurent (2009) emphasise the need for a cultural shift from the top-down, expert-led extension system to a multi-disciplinary approach based on transformative learning, with farmers involved in designing training schemes, and university courses in agricultural extension.

In an interesting reversal of roles, an innovative pilot project in the **United Kingdom** is using experienced hill farmers to train government environmental advisors. The advisers will gain first-hand experience of farming before they provide advice to farmers on setting up individual agri-environment management contracts. It is hoped this training will improve the environmental "fit" with the farming system, and hence the cost-effectiveness and environmental outcomes.[6]

There have been mixed experiences of training initiatives, with several issues arising in their design and delivery (Dwyer et al., 2012; Faure, et al., 2012 and 2011; Laurent, Cerf and Labarthe, 2006). A key question is whether it is possible to reach a target audience by using only voluntary approaches, which may not be attractive to the farmers concerned. Compulsory training may only be effective under certain conditions however.

In **Estonia**, to overcome poor uptake of voluntary agri-environment measures, voluntary training sessions were organised to improve farmer knowledge of environmental land management. The purpose of these workshops was to facilitate mutual self-help among farmers and their continued involvement in environmental land management, for example by providing feedback on the on-going scheme evaluation process. Training workshops also provided opportunities for farmers to share best practice, and for direct two-way communication between farmers and managing authorities.

After noting that advisors have important training needs, some authors (e.g. Ludwig, 2007) in the **United States** show that the initial training of rural development advisors must be revised to better take into account local realities and to improve the skills needed to support farmers in the design and implementation of their own projects.

Höckert and Ljung (2013) analyse efforts to adapt advisory services in **Sweden** to new demands over the past 15 years. The focus was on those efforts that specifically aimed to support farmers to become more competitive and viable so that their farms could become sustainable. The traditional role of advisors as experts was questioned.

Training, particularly for individuals and small groups, can be expensive. Farmers are often unwilling to pay the direct costs of environmental training and in some cases are reluctant to dedicate the time either. Consequently, it is common for these costs to be wholly or partly met from public funds. NGOs may also fund training initiatives. Box 3.5 illustrates two successful training schemes run by NGOs in the **United States** and **Turkey**.

Private agri-environmental training is also available, particularly where a market element is involved as with organic farms. For example, the Research Institute of Organic Agriculture in **Switzerland** offers farmers, for a fee, specialist one-on-one training in organic production systems.[7] It also provides a series of courses and apprenticeships, and supports training by providing information via publications and guides. A similar service for organic farmers is also offered in **Austria** and **Germany**, although the training elements have not yet been developed to the same extent.

Box 3.5. Success of NGO-run training schemes in Nebraska (US) and Turkey

In the **United States** the private *eXtension Foundation* has supported a non-government training initiative delivered by the University of Nebraska Lincoln (UNL), with relatively good uptake and positive feedback from participants. The UNL extension service offered educational one-on-one on-site instructions and two-day long workshops about sustainable water management in agriculture. The protection of water resources is particularly important to Nebraska, which sits on the largest underground aquifer in the world and has had to impose limits on water use on approximately 35% of irrigated agricultural land in the state. It was felt that irrigators needed a better understanding of irrigation principles and the available technology. In 2011, 1 450 producers and consultants (representing 5 million hectares of cropland) attended the training offered by this service. Of these attendees, 300 participants gave feedback on the training: 88% specified that the training had affected the quantity of water they used. The changes in water usage saved a total of USD 8.5 million in 2011 from reduced pumping on the 352 000 hectares participating in the feedback. The total reduction in water pumped was 184 million cubic metres, 18% of Harlan County Reservoir (UNL, 2011).

The *HasNa* distance training programme in Diyarbakir, south-east **Turkey** was funded by an American non-profit organisation that seeks to strengthen capacity on sustainably issues through collaboration and setting up of local partnerships (http://hasna.org/program-countries/turkey/distance-education/). Fruit and vegetable farmers in Diyarbakir lacked the knowledge to cope with diseases and pests affecting their crops and livestock in an effective way without creating substantial environmental costs, despite numerous regional research programmes designed to develop appropriate techniques. In order to facilitate the transfer of information between these two actors, HasNa set up a long distance education programme to channel research carried out by national experts to educate local producers about challenging diseases, pests and climatic conditions impacting agricultural production in the region. The pilot programme was judged a success both in terms of uptake and notably improved knowledge among the farmers. The second phase of this initiative aims to increase uptake and educate a larger number of farmers.

Peer group and co-operative initiatives

Group approaches to agricultural extension have been widely applied for many years and there is a better understanding of their strengths and weaknesses and why they can be so effective. The focus on objectives such as social learning, group development and solidarity, building social capital, collective action and empowerment all help to explain the success of group approaches in many contexts (Garforth, 2011).

Group initiatives encourage a two-way dialogue between facilitators and farmers ("knowledge dialogue"). They allow facilitators to observe and gain insight into what makes farmers most receptive to advice and can provide farmers with the skills and inspiration to translate changed attitudes into changed behaviour.

Local farm networks provide an opportunity to link farm business advice and environmental behaviour change. Evidence shows that farmers are less receptive to advice which has been assembled without giving them the opportunity to be involved thus local farmers' networks positively enhance uptake of advice to farmers (Dwyer et al., 2007). Co-operating and partnering with existing networks, intermediaries and stakeholders could be an effective and efficient way of reaching the farming community. Moreover, community and stakeholder involvement creates cohesion, thereby increasing the likelihood of adoption.

The case study on the evaluation of the **New Zealand** Sustainable Farming Fund shows that its process and approach – by requiring significant community and stakeholder involvement – has created producer and community capability and cohesion. These factors increase the likelihood that the project outcomes will be adopted more widely and that developments will continue (Bell and Yap, 2015)

Landcare in **Australia** is an interesting example of a community based service developed locally that has grown to become a nationwide government funded programme without losing its peer group characteristics or basis within local communities (Box 3.6). Most co-operative and peer group initiatives are on a much smaller scale than Landcare, but there are other examples of national schemes that have developed from a small base within the farming community, e.g. in the **Netherlands** and the Agi-Environmental Group Planning in **Canada**.

Box 3.6. Landcare in Australia: A state funded service of 4 000 peer groups

Landcare began in Australia in the 1980s as an autonomous development of farmer groups concerned about the problem of local land degradation, for example through soil erosion and salinisation, habitat loss, diffuse water pollution and disruption of wetland systems. Landcare now comprises 4 000 groups that undertake local research, analysis and action co-funded by government, business and group members. Whilst Landcare receives considerable support from the national government (a total of approximately AUD 1 billion over the past 18 years) its strength is that it is still co-ordinated and managed at a local level. The bottom-up approach characterised as "of the people, by the people, for the people" is considered to be the key to its success in sustainable land management. Farmers are a key part of Landcare and are always present in discussions and the bulk of participants are volunteers. The main objective is to increase awareness of the problems of land degradation and the need for sustainable land use and management. Landcare provides a wide range of supporting services including: developing regional and local objectives and strategies; research and development of best practices; extension, facilitation and training services for farmers; coordination of funding programmes and institutional partnerships at national, regional and local levels; and monitoring and evaluation. Although Landcare has successfully increased awareness in communities, it is suggested that there is a need for greater consistency between communities, in particular in the development of Landcare strategies and the reconciliation of paid workers and volunteers (Youl et al., 2006).

An example from the Upper Burdekin Rangelands in Queensland illustrates the approach of the Landcare agricultural extension services, offering advice, field days, forums and demonstrations aimed at enhancing both the natural resources and overall productivity and management of farmed properties. Over a four-year period one Landcare facilitator completed 94 individual property projects, engaging 33% of the area's graziers and influencing management on over 1.2 million hectares of land. His work has been a key driver behind the generation of AUD 2.2 million in State and Federal funding which has been matched by AUD 2.3 million of in-kind support from Landcare.

Since Landcare was established awareness and knowledge about sustainable land use and land degradation has increased hugely. By 2007 85% of Australians were aware of at least one aspect of Landcare and 41% of farmers were involved in their Landcare community projects (Department of Agriculture, Fisheries and Forestry, Commonwealth of Australia, 2008). Another mark of Landcare's success is that its community-based approach is being adopted in other OECD countries (**New Zealand**, **United States**, **Canada**, **Iceland** and the **United Kingdom**).

Peer groups may be completely autonomous, but also can be facilitated by public bodies or NGOs as part of an extension service. In **Ireland**, for example, building on the New Zealand experience with Monitor Farms, farmer discussion groups have become an important means of interaction between public sector advisers and their farmer clients (Teagasc, 2008). Currently, Teagasc (Agriculture & Food Development Authority) operates a total of 697 discussion groups that cover dairy, beef, sheep and tillage, with 12 000 farmer members. Since 2009, government policy, through the Department of Agriculture, Food and the Marine, has been supporting an increase in discussion group numbers through the Dairy Efficiency Programme (DEP), Beef Technology Adoption Programme and Sheep Technology Adoption Programme. An independent evaluation on the impact of participation in Teagasc dairy discussion groups revealed that the Dairy Efficiency Programme has been successful in broadening the scope of discussion group membership by attracting farmers with smaller holdings, from less advantaged regions (Teagasc, 2013b). Moreover, it was found that discussion group members are more likely to adopt new technologies and best management practices, and achieved better economic returns than non-member counterparts. Farmer to farmer extension is considered to result in more efficient information dissemination, in terms of numbers and the speed of uptake (Garforth, 1998).[8]

Experience of predominantly local co-operative or peer group initiatives is growing in Europe in particular. These take a variety of forms. In the **Netherlands**, *environmental co-operatives* support groups of applicants for voluntary agri-environment schemes, and this has become a well-established means of delivering partly EU-funded agri-environmental support. Benefits for farmers include reduced transaction costs and a greater awareness of agri-environment issues, and for government these cooperatives provide a single point of contact for dissemination of information and improve the quality of applications. Key factors for success are listed as the need for the initiative to be locally organised, bottom-up, organised by land managers and for it to have a participatory agenda (Franks and McGloin, 2006). A key opportunity factor of this approach is its' capacity to deliver environmental policies at a landscape and water catchment scale.

The Small Farmers' Union in **Spain** established the information campaign *Cambiar el Cambio* (Changing the Change)[9] to increase farmer awareness of climate change in Galicia, North West Spain. Lack of understanding was a hindrance to climate change adaptation by the agricultural and forestry sectors, and was clearly illustrated by a survey at the start of the project to see how rural actors in Galicia viewed climate change issues. For example, although farmers and foresters recognised that pest and disease outbreaks and forest fires had increased in recent years, they did not associate these trends with climate change. The two-year project *Cambiar el Cambio* focussed on the dissemination of information about climate change to farmers. "Ecoguides", brochures, posters, conference and educational materials were circulated. A secondary priority was the facilitation of networking between farmers, agricultural advisors and regional authorities. Following this information campaign, there was an improvement in perception among Galician farmers about adaptation and mitigation measures in agriculture and forestry (Unión Agrarias, 2010).

The *Focus on Nutrients Initiative* in **Sweden** is a peer group initiative to change farmer behaviour as a first step towards meeting Sweden's national Environmental Quality Objectives.[10] The main aim is to make farmers more aware of the importance of zero eutrophication, good-quality groundwater, a non-toxic environment, thriving wetlands and reduced climate impact (Greppa Näringen, 2012). The advisory service is co-ordinated by the Swedish Farmers' Union in collaboration with the Swedish Board of Agriculture, county administrative boards and farm advisory firms. The service offers farmers direct advice on climate issues and plant protection application from advisors, and also an online advisory service that is available at all times. It is targeted at full-time farmers and currently has 9 500 members, covering 34% of arable land in the region. There has been positive feedback from farmers: 75% claimed they have become more environmentally friendly since receiving advice through this scheme, and all participating farms have seen a decline in nutrient surplus, with nitrogen leaching falling by 800 tonnes per annum and phosphorus leaching by 15-30 tonnes per annum (Greppa Näringen, *op. cit.*).

In **England**, public bodies such as Natural England advise on land management and hold regular events where networks of farmers meet. These events attract exhibitors providing technical advice and are another useful way to find and influence farmer networks. Such events typically provide opportunities to provide one-to-one advice. The *Forum for Sustainable Farming* communicates sustainable agriculture standards to growers, encouraging them to adopt best practice advice and share ideas among growers to encourage innovation among member farms.

In addition to being an effective way of delivering advisory, training and extension services, peer groups can also provide a means for land managers to persuade regional and national authorities to improve policies for environmental land management, illustrated by the examples from **Chile** and the **Czech Republic** in Box 3.7.

Box 3.7. Peer group initiatives influence policy changes in Chile and the Czech Republic

Tierra Viva is an initiative by a group of farmers and environmentalists interested in promoting organic agriculture in Chile (www.vivatierra.com/chile/). They have established a network for farmers to share best practice and to support the marketing of organic produce. In addition to providing support for farmers, Tierra Viva successfully lobbied the Chilean government to introduce a national law for organic certification for agricultural produce, approved in 2006.

In the Czech Republic, a group of hunters approached farmers in their area to encourage them to take up a government scheme subsidising local biodiversity management, which they considered important for sustaining local wild species and habitats and, consequently, beneficial for their hunting. The hunters' group became a key source of information about the scheme among farmers, and also established a network of contacts that facilitated uptake of the scheme. One key outcome of this example of mutual 'self-help' between two peer groups is that the government also took notice of the success, and plans to introduce agents to facilitate farmer uptake of such schemes in future.

Measures based on information and communications technology (ICT)

In a rapid changing policy arena farmers, landowners and their advisers share a growing need for day-to-day information on a widening range of issues, such as weather, climate change, biodiversity, agronomic, environmental and climatic conditions, production practices and innovation, land, water and other input use, markets, the economic situation and current policies and regulations (OECD, 2012). Information transfer is a two-way process, benefitting both the public authorities and the participating farmers.

Early warning systems are one way of using ICT to improve farmers' management capacity. There is a growing need for early warning systems in agriculture as the vulnerability of production systems to climate change increases (Hjerp et al., 2011). For example, in the **United States**, in the autumn of 2012, following the worst drought for 25 years, the United States Department of Agriculture (USDA) called upon software developers to design a smart phone "app" that would provide drought-stricken farmers and ranchers with one-click access to the nearest USDA service centres and drought programmes. If the ICT industry can meet the challenge farmers will have easy access to information on: government assistance, differentiated by location and sector; types of loans or refinance options (with a handy repayment calculator and eligibility requirements); drought maps; and localised weather forecasts.[11]

In **New Zealand**, the government established the FarmsOnLine initiative in 2011.[12] This is a government-owned database which brings together existing information on the ownership and management of rural properties, land use, stocks and crops. It provides a hub for rural information that is vital in a disease outbreak such as foot-and-mouth disease or rural emergencies such as floods. Improving the response times in such events can significantly reduce losses to farmers and the country's economy.

There is a growing and steady trend for farmers to use the Internet to source information and advice. In 2006, 42% of farmers in **England** sought advice online and 45% of livestock farmers used the Internet to obtain information about farm inputs, although more dairy farmers used the Internet as a source of advice (49%) as opposed to 42% of sheep farmers who used the Internet for information (NFRU, 2010).

In **France**, the *Institut national de la recherche agronomique* (INRA) has developed the diagnostic tool DIAGNOPLAN which can identify and localise pest diseases in the field. Farmers can send photos via smartphone of their contaminated crops and receive advice from a specialist.[13]

In the **United States**, estimates by the USDA of farm computer usage and ownership shows that 67% of US farms had access to a computer in 2013, up 5 percentage points from 2011. However, only 40% report use a computer for their farm business, with computer use rising the larger the farm (USDA, 2013).[14]

Box 3.8. Government policy encourages broadband Internet access and use in the United States

The provision of US broadband Internet infrastructure and services is largely privately financed. However, because of its perceived economic and social benefits, several public programmes and policies aim to encourage greater investment in rural areas. Federal and state government policy has helped drive increases in broadband availability in rural areas, generally by leveraging private-sector funds.

Government broadband policy falls into two categories: (i) programmes that encourage investment in hardware and software for broadband networks, and (ii) programmes that encourage greater use of the Internet. USDA's Rural Utilities Service (RUS) has been the lead federal agency for increasing broadband access in rural areas primarily through three programmes:

- RUS's traditional rural telecommunication programme for improving or expanding infrastructure. As part of the loan application, RUS requires that communications facilities receiving RUS financial assistance be capable of providing businesses and households with broadband Internet service.
- RUS's broadband loan programme (authorised by periodic Farm Acts), which lent over USD 1 billion to rural providers to build broadband-capable facilities over the last decade.
- The Community Connect Broadband Grant Program, which services rural communities least likely to receive broadband service.

Source: Stenberg, P. (2013), *Rural Broadband At A Glance, 2013 Edition*, USDA/ERS, Economic Brief, No. EB-23, Stenberg et al. (2009), *Broadband Internet's Value for Rural America*, USDA/ERS, Economic Research Report No. ERR-78, http://www.ers.usda.gov/publications/err-economic-research-report/err78.aspx.

ICT initiatives to promote sustainability have also been developed and used by the private sector. An example is the Cool Farm Tool (CFT), originally developed by Unilever and researchers at the University of Aberdeen.[15] This free online, farm-level greenhouse gas emissions calculator helps farmers measure the on-farm carbon footprint of their crop and livestock production. The CFT is based on empirical research from a broad range of published data sets. Unlike many other agricultural greenhouse gas calculators, this tool includes calculations of soil carbon sequestration.

To be effective, information must be accessible to farmers and other land managers. This requires a range of communication tools and techniques to suit individual needs, and ensuring that the "message" is clear and easily understood by the target audience and relevant to their circumstances. Electronic information, particularly that based on GIS technology, is becoming an increasingly important tool for farmers and many now have easy access to detailed, map-based agri-environmental information about their own farms.

Modern technology provides information about areas of natural importance and/or with sensitive soils, data on water catchments, and increase capacity for agri-environmental land management. It can also facilitate monitoring of land management for public authorities and for this reason is often used at a national scale and the information generated is commonly provided by government advisory bodies. A potential problem with the use of modern technology to increase farmer capacity is the unintentional discrimination against rural actors that do not have Internet access.

While evidence suggests that with the growth of the digital economy, more economic activities are taking place on the Internet, and thereby potentially reducing geographic constraints, increasing efficiency, and improving growth prospects for rural communities, these communities may be at a disadvantage in reaping the benefits of this growth because broadband Internet connections (which offer higher speed Internet access than does dial-up connections) are not commonly available, nor used as readily by rural households as by urban households (EU SCAR, 2013; USDA, 2013).

For this reason it is often argued that targeted support for the provision of ICT tools would facilitate access to up-to-date information about markets, policies and weather that is needed to guide producer decisions, and would offer a gateway to specific kinds of extension advice (OECD, 2012).

Map-based information, with its spatial specificity, is particularly relevant for informing and guiding decisions that relate to the environment. The **Czech Republic** provides an Internet-based Land Parcel Identification Service (iLPIS) to farmers free of charge. This service provides the information farmers need to comply with CAP cross-compliance requirements and to apply for payments for agri-environment management. This online service provides detailed map and orthophoto-based information at the scale of individual parcels on, for example:

- Current schemes and areas designated under the CAP and other policies (Less Favoured Areas, Nitrate Vulnerable Zones and wildlife protected areas).
- Landscape features, their type and management responsibility.
- Breeding areas of farmland birds and of semi-natural grassland habitats.
- Recorded nutrient use.
- Suitability of arable land for conversion to grassland.

Maps showing, at field-scale, the local risk of soil erosion are available online as part of the iLPIS service for all farmers to use in applying cross-compliance soil standards and agri-environmental requirements to their land. The iLPIS also offers farmers practical tools to use maps (printing, measuring distance and area, exporting data to GPS, overviewing measures applied to individual parcels) as well as to make additions to maps (e.g. record manure storage, proposed fertiliser use alterations to parcels, and details of crop rotations (Keenleyside et al., 2011).

Canada offers another example. The Agri-Geomatics Service (AGS) developed by the federal department for Agriculture and Agri-Food provides an online national platform for geographic-based information on agriculture and the environment, with the aim is of improving the decision-making process of policy makers and land managers. It is a free service available to all that offers maps, data and interactive tools to plan a sustainable agri-environmental approach to farmland management.[16] It was developed to help understand the agricultural landscape and to monitor and evaluate land use changes. It also acts as a bridge between farm management and policy development, delivery and targeting (Nelson, 2011). AGS is one of four government information services that uses modern technology to build capacity and to inform stakeholders in the Canadian agricultural sector.[17]

Notes

1. This is a key tenet of adoption-diffusion theory (Rogers, 2003), which suggests that access to information is the principle factor leading to the adoption of new technology. It is assumed that once farmers are aware of the options available to them, they will select the option that best fits their farming operation needs and increases productivity (Hooks et al., 1983). Correspondingly, low levels of adoption of sustainable practices have been attributed to a "lack of dissemination of clear and reliable information" (Gamon et al., 1994).
2. Exceptions to this might include situations where another benefit is conditional upon a farmer agreeing to participate in a particular advisory and training measure (e.g. where compulsory training is a requirement for a farmer's contract for agri-environmental incentive payments, or for official certification of an environmental advisor).
3. The major differences concern the funding schemes (e.g. public, private or mixed), the organisations involved (e.g. inclusion or exclusion of the input suppliers) and the advisory methods (e.g. individual, collective, Internet or face-to-face).
4. AIS is based on the National Innovation System (NIS) concept, which is widely used to guide science and technology policy in OECD countries. The NIS emphasises the role of a wide range of factors that influence innovative activity and innovative performance in an economy. In addition to investments in research, such factors include, for example, human resources development and the climate for entrepreneurial behaviour. The AIS concept rejects a linear vision of science that

emphasises the creation of information that is new to the world and then "transferred" to economic agents. From an AIS perspective, the role of agricultural advisory services is to help economic and social agents develop individual and social skills to better identify their constraints or emerging opportunities, to design strategies to address them and to act according to these strategies (OECD, 2013).

5. The Water Evaluation Benchmark sites is also a good example of past efforts of Agriculture and Agri-food **Canada** to link agri-environmental research with technology transfer, involving producers the design and implementation of research initiatives. The Watershed Evaluation of Beneficial Management Practices programme (a nine-year government initiative to determine the economic and water quality impacts of selected agricultural beneficial management practices at nine watershed sites across the country) ended in March 2013.

6. www.foundationforcommonland.org.uk/hill-farming-training-scheme-for-conservation-professionals.

7. Forschungsinstitut für biologischen Landbau, (FiBL) www.fibl.org/en/switzerland/communication-advice/beratung-bildung.html.

8. Although producer organisations are active in numerous OECD countries, studies examining the role of these organisations in the provision of advisory services are scarce.

9. *Cambiar el Cambio* was supported by EU LIFE+ information and communication funding. www.unionsagrarias.org/lifecambiarocambio/resumen.asp.

10. www.greppa.nu/omgreppa/omwebbplatsen/inenglish.4.32b12c7f12940112a7c800022239.html.

11. http://blogs.usda.gov/2012/09/14/usda-drought-code-sprint-giving-americans-one-click-access-to-federal-drought-relief/, accessed 10 October 2012.

12. https://farmsonline.mpi.govt.nz/About/FarmsOnLine.

13. www.inra.fr/Entreprises-Monde-agricole/Resultats-innovation-transfert/Toutes-les-actualites/Di-gnoPlant-R.

14. Large-scale family (i.e. farms with sales and government payments of USD 250 000 or more) reported 84% have access to a computer, 72% are using a computer for their farm business, and 82% have Internet access. Of the moderate-sales farms (farms with sales and government payments between USD 100 000 and USD 249 999), 73% have access to a computer, 56% use a computer for their farm business, and 69% have Internet access. For the low-sales family farms (farms with sales and government payments between USD 10 000 and USD 99 999), 68% reported having computer access, 45% use a computer for their farm business, and 65% have Internet access (USDA, 2013).

15. www.coolfarmtool.org/CoolFarmTool.

16. www4.agr.gc.ca/AAFC-AAC/display-afficher.do?id=1226330737632&lang=eng.

17. The others are: the National Agro-Climate Information Service (NAIS), the Canadian Soil Information Service (CanSIS) and the Earth Observations Service (EOS).

Bibliography

Bell, B. and M. Yap (2015), "The Performance of the Sustainable Farming Fund in New Zealand", Consultant, Report, OECD internal document.

Bright, A., R. Layton and R. Bonney (2008), *A review of the sustainability of Northumberland lamb production in the UK, with particular reference to the environment*, Report commissioned by G. Dixon, www.northumberlandnationalpark.org.uk/__data/assets/pdf_file/0003/144264/sheepfarmingcarbonfootprint.pdf.

Climate Farmers (2012), *Climate Farmers: Good farming practices for climate change mitigation by young European farmers*, http://issuu.com/ekris1989/docs/climatefarmers_english/1.

Department of Agriculture, Fisheries and Forestry, Commonwealth of Australia (2008), *Making a difference: A celebration of Landcare.* www.daff.gov.au/natural-resources/landcare/publications/making_a_difference_a_celebration_of_landcare.

Department for Environment, Food and Rural Affairs (DEFRA), (2002), *Improving access to advice for land managers*, http://randd.defra.gov.uk/Default.aspx?Menu=Menu&Module=More&Location=None&Completed=0&ProjectID=10600#RelatedDocuments.

Dwyer, J., B. Llbery et al. (2012), *How to Improve the Sustainable Competitiveness and Innovation of the EU Agricultural Sector*, Document prepared for the European Parliament, www.inrb.pt/fotos/editor2/inrb/estudo_com_inov_agr_julho_2012.pdf.

Dwyer, J., J. Mills et al. (2007), *Understanding and influencing positive behaviour change in farmers and land managers - a project for DEFRA*, Final report, 30 November 2007, www.inrb.pt/fotos/editor2/inrb/estudo_com_inov_agr_julho_2012.pdf.

European Commission (EC) (2010), *Report from the Commission to the European Parliament and the Council on the application of the Farm Advisory System as defined in Article 12 and 13 of Council Regulation (EC) No 73/2009*, COM(2010) 665 final, European Commission, Brussels.

EU SCAR (2013), *Agricultural Knowledge and Innovation Systems Towards 2020 – An Orientation Paper on Linking Innovation and Research*, Brussels, http://ec.europa.eu/research/bioeconomy/pdf/agricultural-knowledge-innovation-systems-towards-2020_en.pdf.

Eurostat (2013), *Agr-environmental indicator – farmers' training and environmental farm advisory services*, http://epp.eurostat.ec.europa.eu/statistics_explained/index.php/Agri-environmental_indicator_-_farmers%E2%80%99_training_and_environmental_farm_advisory_services.

Faure, G., Y. Desjeux and P. Gasselin (2012), "New Challenges in Agricultural Advisory Services from a Research Perspective: A Literature Review, Synthesis and Research Agenda", *Journal of Agricultural Education and Extension*, Vol. 18, No.5.

Faure G., Y. Desjeux and P. Gasselin (2011), "Revue bibliographique sur les recherches menées dans le monde sur le conseil en agriculture", *Cahiers Agricultures*, Vol. 20, No. 5.

Franks, J. and A. McGloin (2006), "The potential of environmental co-operatives to deliver landscape-scale environmental and rural policy objectives: Lessons for the UK", *Journal of Rural Studies*, Vol. 23, No. 4.

Gamon, J., N. Harrold and J. Creswell (1994), "Educational delivery methods to encourage adoption of sustainable practices", *Journal of Agricultural Extension*, Vol. 35, No. 1.

Garfirth, C., B. Angell et al. (2003), "Fragmentation or creative diversity? Options in the provision of land management advisory services", *Land Use Policy*, Vol. 20, No. 4.

Garforth, C. (2011), *Foresight Project on Global Food and Farming Futures Science review: SR16B Education, training and extension for food producers,* The Government Office for Science, London, www.bis.gov.uk/assets/foresight/docs/food-and-farming/science/11-562-sr16b-education-training-extension-for-food-producers.pdf.

Garforth, C. (1998), *Dissemination pathways for RNR research. Socio-economic Methodologies. Best Practice Guidelines,* Natural Resources Institute, Chatham, UK.

Greppa Näringen (2012) *Focus on nutrients,* http://www.greppa.nu/download/18.5fe620a913671cf1a6b8000953/Focus+on+nutrients+2012.pdf.

Hjerp, P., K. Medarova-Bergstrom et al. (2011), *Cohesion Policy and Sustainable Development,* A report for DG Regio, October, IEEP, London.

Hanson, J. and R. Just (2001), "The potential of transition to paid extension: some guiding principles", *American Journal of Agricultural Economics,* Vol. 83, No. 1.

Höcket, J. and M. Ljung (2013), "Advisory Encounters towards a Sustainable Farm Development - Interaction between Systems and Shared Lifeworlds", *The Journal of Agricultural Education and Extension,* Vol. 19, No. 3, doi: http://dx.doi.org/10.1080/1389224X.2013.782178.

Hooks, G., T. Napier and M. Carter (1983), "Correlates of adoption behaviors: The case of farm technologies", *Rural Sociology,* Vol. 48, No.2.

Keenleyside, C., B. Allen, K. Hart, H. Menadue, V. Stefanova, J. Prazan, I. Herzon. T. Clement, A. Povellato, M. Maciejczak and N. Boatman (2011), *Delivering environmental benefits through entry level agri-environment schemes in the EU,* Report Prepared for DG Environment, Project ENV.B.1/ETU/2010/0035, Institute for European Environmental Policy (IEEP), London. www.ieep.eu/publications/2012/03/delivering-environmental-benefits-through-entry-level-agri-environment-schemes-in-the-eu.

Klerkx, L. and A. Proctor (2013), ''Beyond fragmentation and disconnect: Networks for knowledge exchange in the English land management advisory system'', *Land Use Policy,* Vol. 30, No. 1.

Klerkx, L. and J. Jansen (2010), "Building knowledge systems for sustainable agriculture: supporting private advisors to adequately address sustainable farm management in regular service contacts", *International Journal Agricultural Sustainability,* Vol 8, No. 3.

Klerkx, L. and C. Leeuwis (2009), ''Shaping Collective Functions in Privatized Agricultural Knowledge and Information Systems: The Positioning and Embedding of a Network Broker in the Dutch Sector", *The Journal of Agricultural Education and Extension,* Vol. 5, No. 1.

Knickel, K, G. Brunori, S. Rand and J. Proost (2009), "Towards a Better Conceptual Framework for Innovation Processes in Agriculture and Rural Development: From Linear Models to Systemic Approaches", *Journal of Agricultural Education and Extension,* Vol. 15, No. 2.

Knuth, U. and A. Knierim (2013), "Characteristics of and Challenges for Advisors within a Privatized Extension System", *The Journal of Agricultural Education and Extension,* Vol. 19, No. 37.

Labarthe, P. (2005), "Performance of services: a framework to assess farm extension services", Paper prepared for presentation at the 11th seminar of the EAAE (European Association of Agricultural Economists), *The Future of Rural Europe in the Global Agri-Food System,* Copenhagen, Denmark, 24-27 August, http://ageconsearch.umn.edu/bitstream/24712/1/cp05la01.pdf.

Laurent, C., M. Cerf and P. Labarthe (2006), "Agricultural Extension Services and Market Regulation: Learning from a Comparison of Six EU Countries", *Journal of Agricultural Education and Extension,* Vol. 12, No. 1.

Labarthe, P. and C. Laurent (2013), "Privatization of agricultural extension services in the EU: Towards a lack of adequate knowledge for small-scale farms?" *Food Policy,* Vol. 38.

Labarthe, P., F. Gallouj and C. Laurent (2012), "Privatization of extension services: which consequences for the quality of the evidence produced for the farmers?" 10th International Farming System

Association (IFSA) Symposium, *Producing and Reproducing Farming Systems: New Modes of Organisation for Sustainable Food Systems of Tomorrow*, 1-4 July, Aarhus, Denmark.

Labarthe, P. and C. Laurent (2009), "Transformations of agricultural extension services in the EU: Towards a lack of adequate knowledge for small-scale farms", Paper presented at the 111 EAAE-IAAE seminar Small farms: Decline or persistence, 26-27 June, University of Kent, UK, http://ageconsearch.umn.edu/bitstream/52859/2/103.pdf.

Lobley, M., E. Saratsi and M. Winter (2010), *Training and advice for agri-environmental management. BOU Proceedings – Lowland Farmland Birds III*, www.bou.org.uk/bouproc-net/lfb3/lobleyetal.pdf.

Ludwig B. (2007), "Today is yesterday's future: globalizing in the 21st century", *Journal of International Agricultural and Extension Education*, Vol. 14, No. 3.

National Farm Research Unit (NFRU) (2010), "Farmers tuned in to online communication", www.nfru.co.uk/press/farmers-tuned-in-to-online-communication.html.

Natural England (2009), "Integrating water issues in farm advisory services: The English experience", FAS and Water Workshop, 21 October.

Nelson, S. (2011), *Federal Agricultural Monitoring in Canada.* Presentation for *GEOSS in Americas Symposium*. Santiago, Chile, October 2011. www.geossamericas2011.cl/lectures/AGRIC/AgrculturalMonitoring.pdf.

Øgaard, A. (2011), Improved water quality by cooperation. Main findings from Aquarius - Farmers as water managers in the North Sea Region, www.landbrugsinfo.dk/Miljoe/Filer/pl_12_823_b2_engelsk_rapport.pdf.

OECD (2013), *Agricultural Innovation Systems: A Framework for Analysing the Role of the Government*, OECD Publishing, Paris. DOI: http://dx.doi.org/10.1787/9789264200593-en.

OECD (2012), *Improving Agricultural Knowledge and Innovation Systems - OECD Conference Proceedings*, OECD Publishing, Paris. DOI: http://dx.doi.org/10.1787/9789264167445-en.

Rural Economy and Land Use Programme (RELU) (2011a), *Field advisors as agents of knowledge exchange: investigating the field expertise of specialist advisors and their role as knowledge brokers between scientific research and land management*, Policy and Practice Note No. 30, Centre for Rural Economy, School of Agriculture, Food and Rural Development, Newcastle-upon-Tyne.

RELU (2011b), *Improving the success of agri-environment initiatives: investigating whether agri-environmental schemes could be made more effective and whether training for farmers would help to achieve such improvements.* Policy and Practice Note No. 37, Centre for Rural Economy, School of Agriculture, Food and Rural Development, Newcastle-upon-Tyne.

Remy, J., H. Brives and B. Lemery (2006), *Conseiller en agriculture*, Educagri Editions, Dijon France.

Rogers, E. (2003), *Diffusion of Innovations*, The Free Press, New York.

Slee, B., D. Gibbon and J. Taylor (2006), *Habitus and style of farming in explaining the adoption of environmental sustainability-enhancing behaviour*, DEFRA/University of Gloucestershire.

Stenberg, P. (2013), *Rural Broadband At A Glance, 2013 Edition*, USDA/ERS, Economic Brief, No. EB-23, http://www.ers.usda.gov/publications/eb-economic-brief/eb23.aspx.

Stenberg, P., S. Vogel, J. Cromartie, V. Breneman and D. Brown (2009), "Broadband Internet's Value for Rural America", USDA/ERS, *Economic Research Report* No. ERR-78, www.ers.usda.gov/publications/err-economic-research-report/err78.aspx.

Stubbs, M. (2010), *Technical Assistance for Agriculture Conservation*, Congressional Research Service. www.crs.gov.

Teagasc (2013a), *Knowledge Transfer Conference 2013: Future of Farm Advisory Services, Delivering Innovative Systems*, Dublin, www.teagasc.ie/events/2013/20130612.asp.

Teagasc (2013b), *Impact of Participation in Teagasc Dairy Discussion Groups Evaluation Report,* Teagasc, Agriculture and Rural Development Authority, Dublin, http://www.teagasc.ie/publications/2013/1844/Discussion_Group_Report_Web_Jan2013.pdf.

Teagasc (2008), *Supporting science-based innovation in agriculture and food: Teagasc statement of strategy 2008–2010*, Teagasc, Agriculture and Rural Development Authority, Dublin, http://www.teagasc.ie/publications/2012/1289/Teagasc-Strategy-Statement.pdf.

Unión Agrarias (2010), *Layman Report. Changing the Change: the Galician agriculture and forestry sector facing climate change*. Results and conclusions report. www.unionsagrarias.org/lifecambiarocambio/docs/INFORME_LAYMAN_EN.pdf.

United States Department of Agriculture (USDA) (2013), *Farm Computer Usage and Ownership*, USDA, National Agricultural Statistics Service, August 2013.

University of Nebraska-Lincoln (UNL) (2011), *Agriculture and Water Management*. www.extension.unl.edu/c/document_library/get_file?uuid=daf0939b-074e-48de-af3f-ffcc785e1cb2&groupId=1873&.pdf.

Youl, R., S. Marriott and T. Nabben (2006), *Landcare in Australia Founded on Local Action*, SILC and Rob Youl Consulting Pty Ltd., http://www.agriculture.gov.au/ag-farm-food/natural-resources/landcare/publications/landcare-australia.

Chapter 4

A framework for evaluating green growth initiatives in agriculture

This chapter discusses the conceptual and analytical issues surrounding the evaluation of the performance and impacts of training, advisory services and extension initiatives in the agriculture. The key objective is how these can be best provided to farmers in order to encourage and enable long-term, environmentally beneficial changes in farming practices.

Evaluations of the outcomes, effectiveness and efficiency of the training, advisory services and extension initiative programmes are scarce. Garforth et al. (2002) noted there were few schemes or programmes for which thorough quantitative evaluations had been done, and even fewer evaluations using comparative approaches between schemes or programmes.

This observation remains largely true today. Faure et al. (2012) conclude that evaluation and impacts of advisory services remain a fertile area of study, while OECD (2012) pointed to a continuing lack of data, targets, and systematic evaluation of national schemes, making it difficult to compare performances across countries. Although there have been some studies on the implementation of specific services in individual OECD countries, e.g. AEA (2010), ADE (2009), Labarthe (2005a; 2005b) and Stubbs (2010), most recent work at the international level has focused on developing countries (Waddington et al., 2010; Anderson and Feder, 2007; Birner et al., 2006).

Notwithstanding the scale of public and private investment in many OECD countries, there is so little empirical evidence of the effectiveness of the advisory, training and extension initiatives in supporting the delivery of agri-environmental policies. At a time of increasing pressure to reduce public expenditure in agriculture, it is important to be clear about the overall objectives and the role of these measures within the overall policy mix.

This chapter begins by setting out at a highly conceptualised level the ways in which advisory, training and extension services might foster green growth. It addresses methodological questions related to: i) the evaluation of the *performance* management of these measures in providing relevant, credible and up-to-date advice; and ii) the evaluation of their economic and environmental *impacts*, such as productivity improvement, skills and adoption of environmentally benign farming practices.

A schematic overview depicting a complex system of causal pathways (of different duration) between the agricultural advisory, training and extension services is presented on the one hand, and the various dimensions of green growth on the other. This underlies the difficulties to perform a full evaluation in which *all* aspects of performance and impacts are taken into account. An important policy implication is that policy makers have a choice of how and to what extent to evaluate the impact of advisory services on green growth. Their choice should be based on an awareness of what might be missed or overstated if some causal pathways are overlooked.

Overview of the causal pathways linking agricultural advisory, training and extension services, and green growth objectives

There are major conceptual and methodological difficulties in analysing the performance and impact of agricultural advisory services, training and extension measures to nurture green growth in agriculture. These are associated with several factors, including:

- The difficulty in measuring and tracing the relationship between inputs (such as the type of technical assistance provided or the amount of public expenditures realised) and outputs (such as changes in productivity and environmental impacts). Not only do many factors affect the performance of agriculture in complex and contradictory ways, but these measures are often bundled with other effective mechanisms, such as cost-share payments for adopting new management practices.
- The multiplicity of approaches and increasing diversity and complexity of institutional options in providing and financing these measures.
- The dependence of their success on the broader policy environment.
- Evaluation of intangible services, such as advice, education and training is very difficult because their output is not tangible (PRO AKIS, 2014; Gadrey, 2000).

In principle, the effectiveness of these measures can be evaluated against the achievements of the set targets and desired outcomes. From a policy perspective, the ultimate criterion to assess agricultural advisory services, training and extension measures in fostering green growth is the extent to which they are successful in achieving the stated policy objectives. This will depend on the extent to which these measures have an influence on decision-making at the farm household level and which result in changes to existing practices, e.g. by improving farm management, promoting the adoption of new technologies, and/or fostering innovative behaviour.

Figure 4.1 provides a stylised picture of the various ways in which agricultural advisory services might impact on different dimensions of green growth. The framework, which is based on the framework developed by International Food Policy Research Institute (IFPRI) (Box 4.1), suggests an impact-chain approach to analyse the impact of advisory services, training and extension measures. In this impact-chain approach, *inputs* (quantity and quality) of the advisory, training and extension services lead to *immediate outcomes* (changes in farm households' behaviour) and, finally, to *ultimate outcome impact* (contribution to broader green growth goals).

More specifically, the following key pathways can be distinguished. First, to understand the contribution of agricultural advisory services, training and extension measures to green growth in agriculture, it is essential to consider these measures as part of the wider systems in which knowledge and innovations are generated, disseminated and utilised in the agricultural sector. Agricultural research, advice, training and extension are interconnected and a wide range of institutions and stakeholders are involved in the production and delivery of these measures. Evaluations should take into account all actors involved in their provision and assess the role they play with the wider AKIS (Faure et al., 2012; Birner et al., 2009; Hoffmann et al., 2009; Scoones and Thompson, 2009).

Second, within the wider AKIS perspective, the various factors influencing the supply (quantity and quality) of agricultural advisory, training and extension services, such as institutional structures, the suppliers of advisory services and the method by which advice is provided, play a critical role (Birner et al., 2006; 2009).

Box 4.1. IFPRI's framework for designing and analysing agricultural advisory services

IFPRI has developed an analytical framework to analyse the performance and impact of agricultural advisory services (Birner et al., 2006; 2009). The framework can help policy makers to identify "best fit" policy reform options for financing and providing these services and it can also guide research projects aimed at creating empirical evidence on what works where and why.

The framework focuses on: i) "disentangling" the design elements of advisory services – that is, governance structures, capacity and management, and advisory methods – and their comparative advantages and disadvantages under different frame conditions; ii) performance measurement and quality management in the provision of agricultural advisory services; and iii) impact assessment with regard to multiple goals as well as assessment of the costs and benefits associated with different ways of providing and financing agricultural advisory services.

Four sets of conditions need to be considered when deciding on these characteristics: i) the policy environment; ii) the capacity of potential service providers; iii) the type of farming systems and the market access of farm households; and iv) the nature of the local communities, including their ability to co-operate.

The framework shows that reforms of agricultural advisory services can combine different reform elements – such as decentralisation, contracting-out, using new advisory methods, and changing the management style – in different ways so as to best fit local circumstances. By distinguishing among the various factors influencing agricultural advisory services, the factors contributing to the final impact could be identified. Analysing agricultural advisory service systems, Birner et al. (2006; 2009) suggest it is no longer possible to think in terms of ideal solutions that are applicable everywhere; attention should be paid instead to adapting advisory services to given and diverse situations.

Source: Birner, et al. (2009), "From best practice to best fit: a framework for designing and analysing agricultural advisory services worldwide", *Journal of Agricultural Education and Extension*, Vol. 15, No. 4.

Figure 4.1. Conceptual framework for analysing performance and impacts

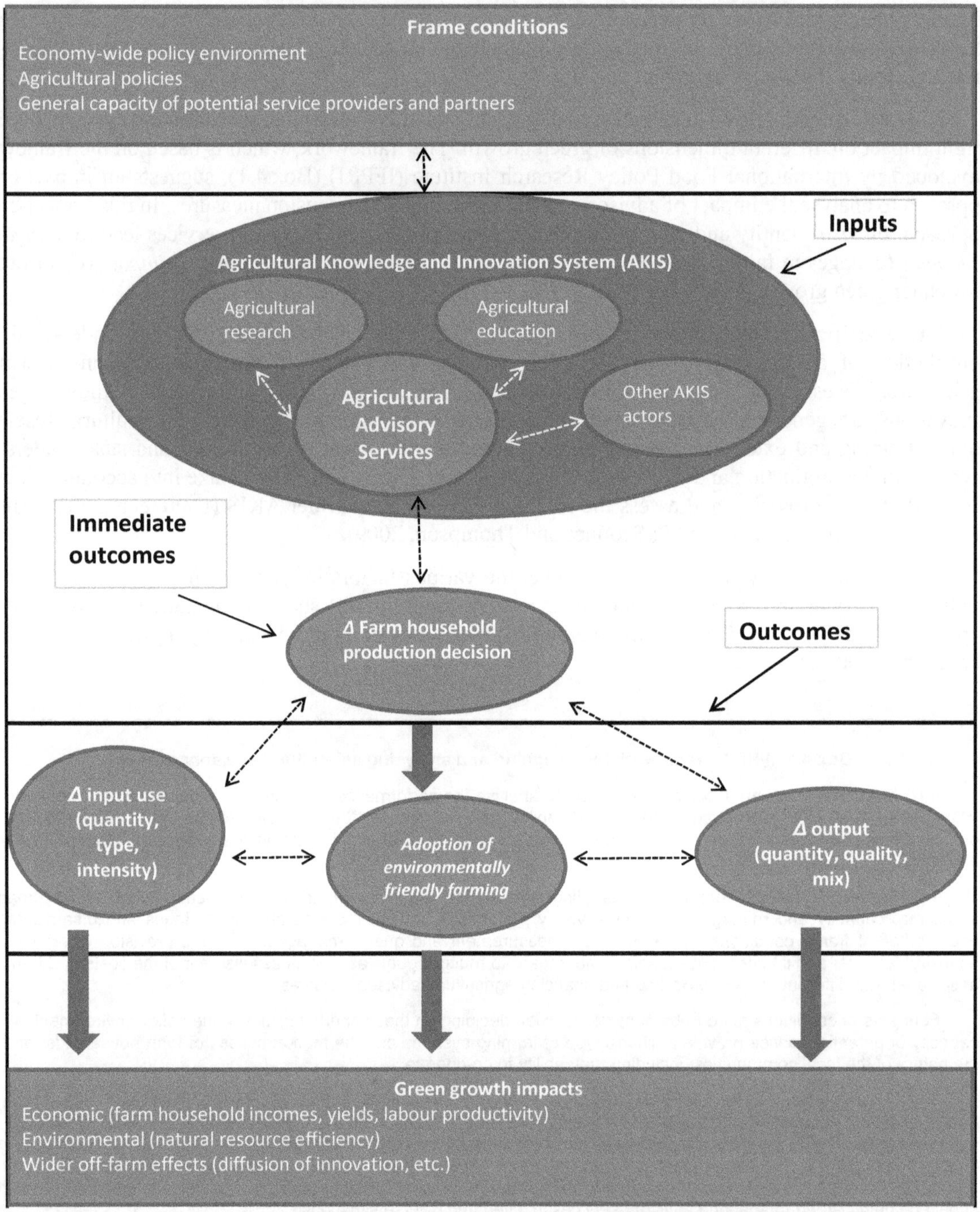

Source. OECD Secretariat; adapted from Birner et al. (2006), *From "best practice" to "best fit": A framework for designing and analyzing pluralistic agricultural advisory services worldwide*, IFPRI, Washington, DC.

A wide range of institutions are now involved in the provision and financing of these measures, including public sector agencies as well as non-governmental organisations and the private sector (Chapter 2). Different providers fulfil different roles, address different issues, and thus provide different types of advice in different ways. Co-ordination between the different advisory service providers and users of the innovation system is crucial to ensure that advice, training and extension is accessible to all, and that they are reliable, relevant and up-to-date (see Tchuisseu and Labarthe, 2015; Faure et al., 2012; Labarthe and Laurent, 2013).[1]

Using advisory measures, training and extension initiatives to support agri-environmental policy agenda means a considerable broadening of focus and an increase in complexity as compared to the conventional model of predominantly "top-down" agricultural extension. The new model is characterised by multiple influences and two-way information flows (Box 4.2).[2]

Third, as indicated by the arrows in Figure 4.1, any impact these measures can achieve with regard to green growth objectives depends on how consumers (i.e. farmers and land managers) make use of the services provided. That is, any impact depends on the extent to which these measures have an influence on decision-making at the farm household level and lead to a change in existing farming practices; e.g. by improving farm management, promoting the adoption of new technologies, and/or fostering innovative behaviour relative to a situation with no policy. As advisory, training and extension measures are usually voluntary, a principal set of questions should concern the rate of uptake of the measure, and whether the producers adopting the measure are those likely to obtain the greater impact from it. The reasons that explain the choices of adopters and non-adopters should also be revealed.

Box 4.2. Theoretical approaches to advice dissemination

The last 20 years have seen major theoretical debate and changes in thinking about the provision of advice to farmers. Agricultural extension theory and methodology has traditionally been predicated on the knowledge transfer model, where extension aims to promote, through dissemination of information and technical solutions, adoption of predetermined practices. However, a changing context has contributed to a paradigm shift with the formulation of alternative models which emphasise human development, where extension aims to facilitate interaction and innovation by the users.

The knowledge transfer approach

The knowledge (or technology) transfer extension approach views knowledge as a discrete, tangible entity which can be transferred between actors. It assumes that innovations (and knowledge) originate in science and are transferred to land managers who adopt them. This transfer follows a linear and sequential "one-way" path, as described by a number of commentators (Röling, 1990; Ruttan, 1996; Black, 2000). Within this perspective, early empirical approaches sought to discover patterns or predictive factors in the way decisions are made on the basis of farmer socio-economic factors, and provision of information. Farmers were categorised as "laggards" or "innovators" according to how readily they adopted innovations. Knowledge transfer was the dominant paradigm and captured the concerns of the so-called "productivist" era of the 1970s and 1980s, in describing the translation of science in order to encourage and promote efficient production.

The knowledge transfer approach has been criticised on three main grounds (Buttel, 2001): i) the approach is no longer appropriate to address the challenges facing modern agriculture, such as sustainability, climate change and the provision of public goods; ii) it does not reflect the empirical evidence of how farmers use information; and iii) it takes no account of other influences upon the uptake of information and advice, including the economic imperatives that drive decisions.

The human development and knowledge exchange approach

The human development approach considers that knowledge is socially constructed through interaction and experience. Communication within a social system or a group is regarded as an important process in articulating, sharing and exchanging ideas among land managers. Theories drawn from knowledge networking; social networking; social movements; social learning; experiential learning; social capital; and systems research underpinned much of the research undertaken which seeks to understand collective behaviour. The role of extension in facilitating collective processes is seen to be critical.

Although seen as an improvement of the knowledge transfer model, there have been a number of criticisms of human development models and methodologies, including: i) lack of a coherent theoretical foundation; ii) issues of legitimacy, accountability and representation are overlooked; iii) problems associated with poor participation practices; iv) challenges in working with conflicting knowledge from multiple sources; and v) the economic drivers of behaviour are not sufficiently taken into account (CCRI/MLURI, 2006).

In contrast to other measures, such as capital grants and agri-environmental schemes, these measures work primarily at the level of seeking to change attitudes and understanding as a means to engender changes in practice, rather than the other way around. Funding advice, training or information may not immediately result in farm changes on the ground in terms of practices or systems. The ways in which the wider policy environment either supports or works against an effective linkage between attitude/understanding and changed practice therefore becomes an important consideration when seeking to assess these measures' effectiveness: a weak conducing policy environment, or one which tends to undermine farmers' capacity to follow through, is likely to reduce the impact on the ground of advisory, extension and training initiatives, whilst a broader supportive context which builds farmers' confidence to act, or provides frequent opportunities for reflection and discussion, can help to turn intentions into behaviours.

Once advice has been sought and/or received, the farmer will decide on whether to change his/her practice. The farmer's willingness to change is an important factor influencing decisions and achieving long–term behavioural change. As shown in Box 4.3, the uptake of advisory services and the adoption of farming practices with potential to foster green growth are influenced by several factors, including the characteristics of the farm households as well as the characteristics of the advisory service and farm practices, and it cannot be taken for granted that farm households will react (or not react) according to the stated policy objectives.

Box 4.3. Factors influencing environmental behavioural change

The literature shows that sociological, cultural, economic, educational and organisational factors can all influence environmental awareness and response to environmental issues. The behaviour change theories and education models identify those factors likely to be most influential in fostering behavioural changes in relation to sustainable land management. The success of non-regulatory or voluntary adoption measures is influenced by a variety of social, psychological and economic factors. These include the following.

- Economic benefits.
- Compliance with environmental legislation.
- Sound practical scientific knowledge of the issues and the effects of farm practices.
- Practical local knowledge and experience.
- Land manager awareness of the environmental issues around farm practices.
- The visibility of the environmental damage.
- The complexity of the proposed innovation.
- The extent to which farmers are engaged in defining problems, identifying solutions and in monitoring progress – successful processes foster interactivity between scientists, technical advisors and land managers.
- The quality of the information provided to farmers – it should be targeted to the needs of the farmers and written in user-friendly language.
- The ability to demonstrate results and the ease with which success can be measured.
- The extent and quality of on-going support to reinforce the decision to change, and build confidence and capacity.
- The personal attitudes of the land manager – those who have a sense of community responsibility and environmental stewardship are more open to adopting environmentally sustainable practices which may involve more time and risk.
- The financial security and level of risk-adversity of the land manager.
- The time pressures on the land manager.
- The level of support among the farmer's social and professional networks for behaviour that respects the environment.
- The strength of networks between farmers, which can be used to facilitate knowledge and awareness and provide mutual support and encouragement.
- The extent to which central government and local authorities lead by example.

Source: Pannell et al., 2006; DEFRA, 2010 and 2007.

Moreover, there is a cost to farmers in time and money spent improving their knowledge base which may not be compensated by higher returns (Charatsary et al., 2011). Pannell et al. (2006) conclude that the adoption of new farm practices takes places when the farm practice helps to attain farmers' objectives, such as economic, environmental or social goals.[3]

The likelihood of bringing about positive environmental behaviour change largely depends on the extent to which the users are convinced that: i) the environmental issue to be addressed is serious; ii) it affects them; iii) the measures will solve the problem; and iv) they are capable of performing the recommendations (Roger, 2003).[4] This will require greater partnership working between farmers and researchers and that the farmers accept that there is a problem and moreover, that they have a responsibility for it.

The ability of farmers to formulate demand is also of crucial importance to the performance of agricultural advisory, training and extension services. This ability is influenced by the characteristics of the farm households and the socio-economic environment in which they live as well as by the characteristics of the measures themselves. For example, a decentralised governance structure, a favourable advisory staff to farmer ratio and the use of participatory advisory methods are all factors that improve the possibilities of farm households to express their demand and hold the service providers accountable.

Fourth, from an AKIS perspective, the performance and impacts of agricultural advisory services also depend on *contextual* (*frame*) factors. These might include factors such as the overall innovation policy environment, the overall policy stance of a country's policies towards advisory, training and extension measures, the level and type of the agricultural support and agri-environmental policies in place, the socio-economic structure of the sector (education levels, age), structure of production (e.g. crops, livestock), farm structures and agro-ecological potential.

Evaluating performance of advisory, training and extension services

Assessments of the implementation performance of advisory, training and extension services to support agri-environmental management can be undertaken by defining criteria referring to effectiveness (achievement of objectives), efficiency (results obtained compared with resources invested), quality of services provided and equity of access to services.

Birner et al. (2009) show that three main components of an advisory system interact and explain its performance: i) the governance structure, including financing mechanisms and relationships between partners; ii) the method by which advice is provided and; iii) the capacities of advisory service providers (the organisations providing advice), including their management approach and the characteristics of individual advisors.

A wide range of implementation performance indicators that capture the supply and quality of advisory, training and extension services may include: i) the accuracy and relevance of the contents of the advice; ii) the timeliness and outreach of the advice, including the ability to reach disadvantaged groups (e.g. number and share of farmers having made use of environmental farm advisory services; number of farmers visited; number of farmers per adviser; number of training days received); iii) the quality of the partnerships established and the feedback effects created (e.g. time and frequency of the provision of the advisory services to the farm); and iv) the efficiency of service delivery. The relative importance of these indicators depends on the policy objectives of the measures, and there may be trade-offs among them.

From an analytical perspective, measuring and explaining the implementation performance of advisory, training and extension providers is less demanding than assessing impact because performance can be attributed in a more straightforward way to the characteristics of the providers of these measures. For example, changes in spending can be linked directly to changes in the number of farmers reached, but it cannot be linked directly to farm household incomes or environmental outcomes.

From a policy perspective, however, it is the impact (outcome) of these measures in terms of their contribution to green growth goals that is ultimately important (see next section). But, analysis of implementation performance is useful not only because it provides an important tool for improving these measures, but also because it provides useful insights for achieving outcomes. For example, specified targets such as uptake of the measures (e.g. area of land adopting a certain management practice) could be directly linked to environmental outcomes.

In the literature, several methods and approaches – including econometric techniques and use of experimental design – are used to overcome methodological challenges in measuring the performance of these measures, with varying strengths and weaknesses. The results of these assessments are mixed and depend on the context, the methods chosen and the research question addressed by the authors (Faure et al., 2012).

The costs of advisory services are relatively easy to assess as long as the advisory services are publicly financed and transaction costs are not taken into account. Concern over data quality, along with difficult methodological issues regarding causality and quantification of all benefits, must be important qualifiers to the prevailing evidence of good economic returns from extension.

Comparing different advisory systems raises formidable methodological questions, notably when statistical tools are used. The rare studies that exist use more qualitative approaches, such as that of Labarthe (2005), who draws on institutional economy to compare the case of the **Netherlands**, **Germany** and **France**. In the **United States**, Lohr and Park (2003) use an econometric model to explain the differences observed between different advisory sources based on surveys of organic farmers. Studies using transaction cost economics to evaluate the performance of an advisory service system are uncommon, although such assessments would be necessary to help formulate policy recommendations (Faure et al., 2012; Bartoli and Rocca, 2012; Charatsary et al., 2011; Birner et al., 2006).

Comprehensive evidence on the number of advisors per farmer and on the uptake of advisory, training and extension measure to support environmental management is not available across the OECD countries. In the **European Union**, the evaluation of the FAS shows that, for the countries for which data is available, the average outreach for 2008 of on-farm one-to-one advice as share of farmers that received direct payments is 4.8% (ADE, 2009). In **Greece**, there is one advisor available for every 862 farm holdings that receive direct payments (909 230 holdings); for holdings that receive more than EUR 10 000 per year (130 950 holdings) this ratio is 1 to 124 (Damianos, 2015).

In the **EU-27**, the total number of participants in vocational training and information actions specifically devoted to the maintenance of landscape and the protection of the environment, represented 16% in 2010 (or nearly 164 000) of the total participants in vocational training and information actions supported under the rural development measure 111 (Vocational training and information actions) (Figure 4.2). The total public spending for this measure was EUR 4.2 million (or, on average, EUR 330 per participant), with **Italy** accounted for about a third (32%), followed by **Belgium** (23%) and **Hungary** (22%) (Eurostat, 2013).

At the **EU-27** level, 8.5% of the advisory services supported by rural development policy (rural development measure 114) in 2010 were specifically devoted to the environment. In **Belgium**, the **United Kingdom**, the **Czech Republic**, **Hungary**, **Italy** and **Poland** this share is even higher, particularly in Belgium and the **United Kingdom** where it is 86% and 48% respectively) (Figure 4.2). Total public spending for this measure was EUR 53 million (or, on average, EUR 590 per application), with **Sweden** accounted for more than half (64% or EUR 34 million) (Eurostat, 2013).

Figure 4.2. Farmers' training and use of advisory services devoted to the environment, OECD EU member countries, 2010

Participants in environmental training[1]

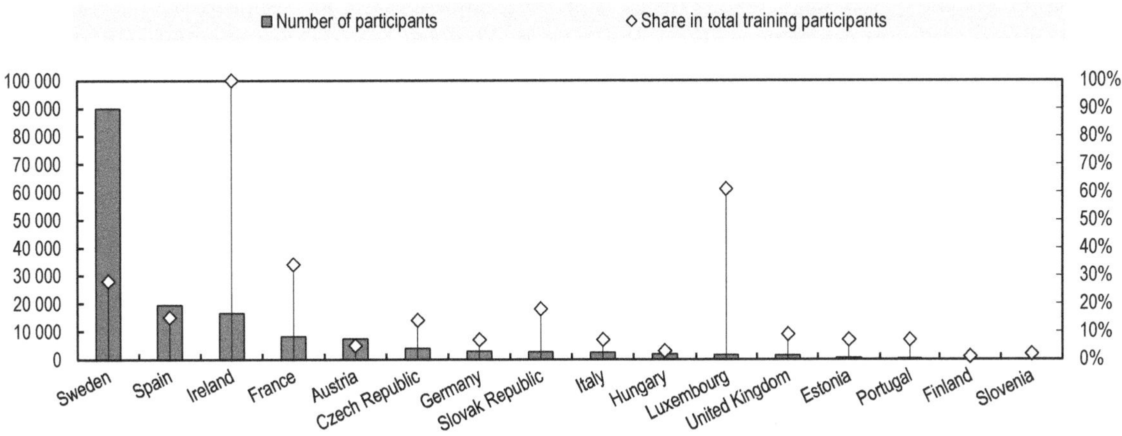

Applications to environmental advisory services[2]

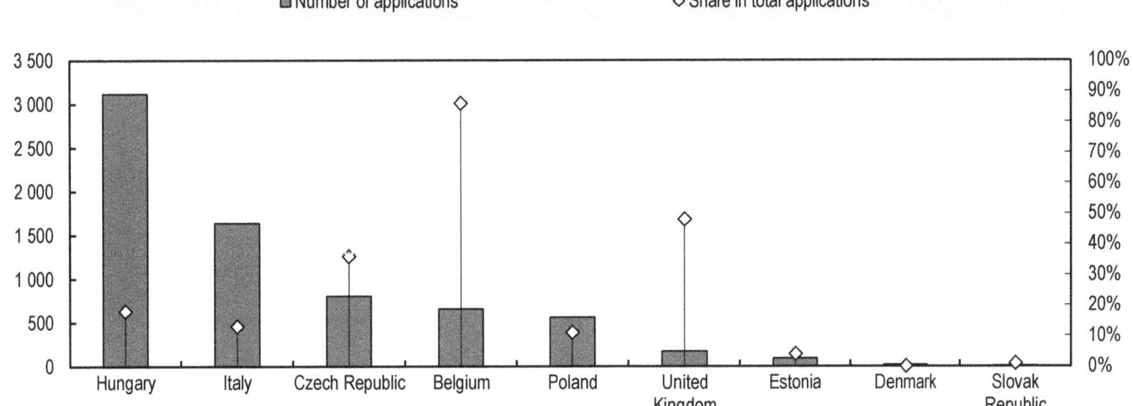

Economic actors benefitting environmental training and information[3]

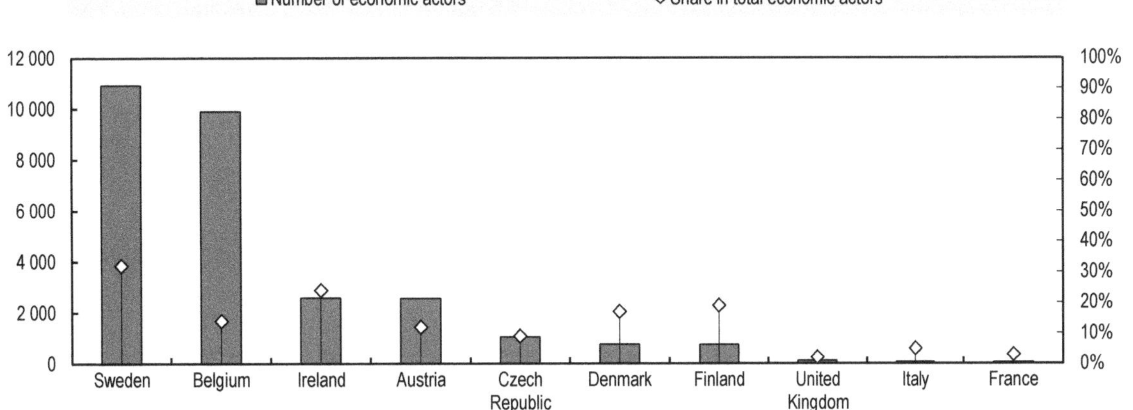

Note. Background data are based on rural development data, which may be reported differently by different EU member states. Double counting may also be present in the total numbers of participants, as is the case for example in Sweden. Differences in the design of individual training activities (e.g. their length) also contribute to the variation in number of participants. 1. Rural development measure 111; 2. Rural development measure 114 ; 3. Rural development measure 331.

Source: Eurostat (2013), Agri-environmental indicator – farmers' training and environmental farm advisory services, http://epp.eurostat.ec.europa.eu/statistics_explained/index.php/Agri-environmental_indicator_-_farmers%E2%80%99_training_and_environmental_farm_advisory_services.

The number of rural area economic actors supported to follow training or information actions in the fields of maintenance and enhancement of landscape and protection of the environment (rural development measure 331) was almost 29 000 in the EU-27 in 2010. At EU-27 level, training and information actions in the fields of landscape and environment represent, in terms of the number of economic actors, 17% of the total training and information actions supported through this measure. This share is even higher in **Sweden**, **Ireland**, **Germany** and **Finland** with values between 19% and 32% (Figure 4.2). Total public spending for this measure was EUR 10.3 million in 2010 (or, on average, EUR 358 per economic actor supported), with **Austria** accounted for just over half (52%), followed by **Sweden** (27%) (Eurostat, 2013).

Evaluating outcomes (impacts) of advisory, training and extension measures

From the green growth perspective, evaluation of the outcomes (impacts) of these measures primarily entails assessing the extent to which these measures: i) have improved labour productivity, farm household incomes, and overall competitiveness of the agricultural sector; ii) have contributed to improving natural resource efficiency, particularly for land and water; and iii) have produced wider economic and environmental off-farm effects (e.g. diffusion of innovation).

Assessing the impact of agricultural advisory services involves a range of methodological challenges, which have been widely discussed in the literature (see Birner et al., 2006, for a review). The challenges derive from problems such as multiple goals, attribution problems, measurement issues, lagged impacts, spillover effects, data problems and sample bias. Many of these challenges are also faced when evaluating other policy areas.[5]

Attributing the change or outcome delivered by these measures – which would not have occurred in their absence – is generally very challenging. First, farmers' decisions and performance are influenced by many other systematic and random effects and thus fairly sophisticated methods (e.g. econometric and experimental methods) are required in order to reliably ascertain the impact of advisory, training and extension measures to farmers requires (Birner et al., 2006). A key finding from the behavioural literature is that decision making processes are highly contingent, which increases the challenge of assessing how advisory, training and extension services impact on farming decisions (Kahneman, 2011). This difficulty is especially likely to be more prevalent in situations where non-pecuniary aspects of farming are paramount (such as about agri-environment issues connected to land stewardship, community involvement, family succession of farm ownership), thereby hindering accurate quantifications.

Second, advisory services, training and extension measures to support agri-environmental management often have multiple goals and provide multiple benefits and are usually linked to an existing set of regulations or incentives. For example, in the **United States**, while expenditure data on Conservation Technical Assistance (CTA) are available, measuring the effects of the programme on fostering green growth would be difficult to achieve because the programme has multiple benefits which are often bundled with other policies, such as cost-share payments for instituting new management practices.[6] Likewise, in the **EU**, advisory services under the FAS system are provided to farmers who are eligible for direct payments.

In addition to the difficulty in attributing impact, *measuring* the actual environmental outcome (i.e. improved water quality, reduced greenhouse gases emissions) is very complicated and harder still to assign a monetary value. The impacts of these measures are spilled over to non-participants. For example, information provided by advisory services may be shared with non-participants, causing changes in yields or in use of natural resources for non-participants, as well as participants. Neglecting such effects may produce biased results (either overestimate or underestimate total impacts of these measures). Accounting for such impacts, however, requires rather sophisticated modelling techniques.

Moreover, the impacts of these measures may not occur instantaneously. The full impact of these measures cannot be properly assessed until they have been in place sufficiently long for the farm household's decisions to take effect, and for their consequences to work through to the relevant performance indicator (e.g. increase natural resource efficiency and productivity). It is possible, for example, to observe a shift towards sustainable management practices, but without any immediate environmental outcome.[7]

There are also data problems, including difficulties in implementation and measuring appropriate indicators of inputs, outputs and outcomes; issues of data comparability (especially when different survey instruments are used at different points in time or for different sub-samples); and problems in mixing data from different sources and of different types. Finally, there is the issue of *selection bias* (when the chosen level of advisory, training and extension measures is linked to factors that themselves affect the behaviour of the target variable) or *sampling bias* (i.e. non-random sampling) or *attrition bias* (i.e. drop-out rates) or *confirmation bias* (i.e. a search for confirmatory evidence not to disprove a hypothesis) and even absence of experimental control groups (i.e. farmers not receiving advice), which could also seriously distort the evaluation results.

These difficulties have led some authors to go so far as to argue that attributing and measuring impacts to particular agricultural research and dissemination efforts is futile (e.g. impacts depend on many external factors that cannot be controlled) or too costly and the objectives of the beneficiaries and the objectives assumed by those designing advisory, training and extension services are different (Ekboir, 2003; EIARD, 2003).

A common gap in most research is that quantitative methods are seldom deployed to measure impacts, not least because of the complexity – with both temporal and conceptual challenges – involved (Faure et al. 2012). The review of the literature in the **United Kingdom** points outs that significant gaps exist, some of which are due to the fact that most evaluations focus on individual projects or initiatives with relatively short life spans and/or very small samplings. But even where these measures have been evaluated, only a proportion of their impact will have been captured (e.g. project goals and targets) (Dwyer and Reed, 2015).

A common thread in much of the recent literature is the conclusion that there is no single best method to use in all circumstances. There are many different methods with different strengths and weaknesses that depend on various assumptions (Box 4.4). A combination of qualitative and quantitative methods, using different approaches, could be necessary to produce robust and policy-relevant results.

Box 4.4. Methodological approaches to evaluate advisory services within an AKIS perspective

Different methodologies and analytical approaches have been used in the literature to analyse the performance and impact of advisory services, including econometric techniques, social network analysis; transactions cost analysis; and game theoretic modelling (Spielman, 2005). More recently, the use of experimental design has gained increasing importance in impact evaluation, even though applications to agricultural advisory services are still rare. These approaches have different strengths and weaknesses in overcoming methodological challenges which have been extensively discussed in the literature.

A further methodological approach, developed in the innovation systems context, are innovation indicators which can be used to assess the innovative performance of a sector or country as compared to other countries, and to track changes over time. This approach has been widely used to benchmark the performance of innovation systems in OECD countries. Based on this approach, one can develop a variety of agricultural innovation indicators, including indicators describing capacities and levels of investment in agricultural research, advisory services and training and education; indicators of linkages and partnerships with other actors in agricultural innovation systems; indicators of transboundary technology exchange; indicators of the success of targeting the innovation potential specific group of farmers (e.g. small); and outcome indicators, such as number of new varieties registered, and adoption rates of innovative agricultural practices. Such agricultural innovation indicators, which are suitable for benchmarking across countries, could play an important role in guiding agricultural innovation policies.

The OECD, in the context of its work on agricultural innovation systems, has developed a framework to analyse the role of government in fostering innovation in the agricultural and agri-food sectors. This framework consists of a questionnaire and suggestions for possible benchmarking indicators for agricultural innovations (OECD, 2013).

As in the case of the performance indicators, the indicators used to measure impact depend on the objectives to which these measures are expected to contribute, such as economic growth and environmental sustainability, empowerment and promotion of innovations. Several studies address the evaluation of the impacts of advisory services, although most impact measurements focus on a limited number of criteria, often quantitative in nature, such as changes in agricultural practices (e.g. for integrated pest management), crop yield variations and variation of farm revenues (Akobundu et al., 2004, in the **United States**; Marsh et al., 2004, in **Australia**).

Another approach is "self-evaluation" of farmers who participate in government advisory and extension activities. Such an approach is more in line with the "human development" approach outlined in Box 4.2 (SOLINSA, 2014). In **Sweden**, for example, evaluation of advisory services within the rural development programme is based on the farmers' own assessment of the impact of the advice, rather than field-studies of their behaviour (Swedish Board of Agriculture, 2010; 2011). Results are compiled from a written questionnaire to 8 000 farmers who either received on-farm advice or took part in group-training in 2010 show that 20% of the farmers who received on-farm advice say they subsequently have changed their practices in a more environmentally friendly direction. Concerning group training, the length of the training does not seem to be directly related to the degree of change in behaviour. About 40% of participants say they changed their practices towards an environmental direction. About 10% assess the change as very or quite high.

Some authors make the case for methods that combine quantitative and qualitative analyses, which they esteem are more capable of grasping the complexity of the dynamics (Faure et al., 2012). Although such evaluations are complicated to conduct and expensive to carry out, they are able to establish a causal link between the impacts measured and the advisory service methods implemented.

Research on impacts that go beyond the farm performance level is rare. While some studies aim to describe the learning process in a given advisory situation, few can claim that they aim to measure the impact of the organisation of an advisory service on these processes by highlighting a cause-and-effect relationship.

A classic approach consists of questioning beneficiaries on their acquired knowledge as Hall et al. (2004) did in the **United States** with cattle farmers. In **Australia**, Cameron and Chamala (2004) followed an "Action Research Approach" to assess the impact of an advisory service aiming to enhance the managerial skills of farmers, by using synthetic indicators informed by data collected from farmers in the programme.[8] Other authors seek to establish a link between learning processes and the impacts of these learning processes. For example, King et al. (2001) show that, in **Australia**, a participatory learning method (Participatory Action Learning), in which farmers analyse the lessons of their own personal experiences, improves the effectiveness of individual and collective learning.

Bartoli and Rocca (2012) analyse the attitude of **Italian** farms in gaining access to agricultural extension services, using the AKAP (Awareness, Knowledge, Adoption, Productivity) sequence approach explained by Evenson (1996). Their results confirm that, due to a set of socio-economic constraints which impedes full adoption of agricultural extension services, simple awareness does not automatically engender adoption.[9]

Notes

1. Ludwig (2007) in the **United States** and Marsh and Pannell (2000) in **Australia** demonstrate the importance of generating interactions between research and advisory service activities, notably to facilitate feedback from farmers to researchers regarding the farmers' experience.

2. Labarthe (2005) proposes an analytic framework inspired by service economics that allows a description of the production of innovations within advisory service organisations. He observes five types of innovation involving advisory services, with a focus on i) skills of advisors; ii) service provision methods; iii) information processing; iv) knowledge production and management and; v) the interpersonal aspects of the advisor-client relationship.

3. A meta-analysis by Prokopy et al. (2008) on the determinants of agricultural best management practice (BMP) adoption in the **United States** shows that education levels, capital, income, farm size, access to information, positive environmental attitudes, environmental awareness and utilisation of social networks are the main variables which are generally positively associated with adoption rates. In **Canada**, farmers with an Environmental Farm Plan cited economic and time constraints (55.5% and 23.3% respectively) as their principal reason for not implementing BMPs, while only 6% cite lack of information (STC, 2011).

4. Roger's (2003) protection motivation theory proposes that threatening messages will be effective when they convince recipients that (a) the problem is serious, (b) they are susceptible to the problem, (c) recommendations will alleviate the problem, and (d) they are capable of performing the recommendation. However, the objective should not be to frighten the recipient as, in addition to the moral issue, overstating the case is likely to reduce message effectiveness.

5. See, for example, OECD (2009).

6. The CTA ensures that landowners are made aware of government conservation requirements. It provides financial support to assist them in designing, adopting and employing an appropriate suite of technologies and practices; it helps evaluate the health of the ecological and biological resources; and assists public and private efforts to develop watershed and area-wide conservation plans.

7. This was the case in **New Zealand**, where evaluation of a project aimed at improving economic and environmental outcomes in the Waituna catchment, Southland, conducted from 2005 to 2007, found that the project achieved a noticeable shift towards sustainable management practices. However, a review in 2009 revealed no discernible improvement in water quality (Nimmo-Bell, 2009).

8. Action research is a method of inquiry that entails systematic engagement with a problem or issue in order to find – through experimentation with different approaches – the most suitable way to proceed (www.web.ca/~robrien/papers/arfinal.html).

9. They found that, without any reference to farms' socioeconomic traits, there is high awareness about the existence of agricultural extension services among the majority of the investigated farms. However, knowledge and adoption are strongly reduced to one-third of total potential demand.

Bibliography

ADE (Aide à la *décision* économique) (2009), *Evaluation of the Implementation of the Farm Advisory System*, ADE in collaboration with ADAS, Agrotec and Evaluators EU. Belgium. http://ec.europa.eu/agriculture/eval/reports/fas/index_en.htm.

AEA (2010), *Agricultural advisory services analysis*, Report to DEFRA, Issue 4, July 2010, http://archive.defra.gov.uk/foodfarm/landmanage/climate/documents/advisory-analysis.pdf.

Ahnstrom, J, C. Francis et al. (2009), "Farmers and nature conservation: What is known about attitudes, context factors and actions affecting conservation?" *Renewable Agriculture and Food Systems*, Vol. 24, No.1.

Akobundu, E., J. Alwang, et al. (2004), "Does Extension Work? Impacts of a Program to Assist Limited-Resource Farmers in Virginia", *Review of Agricultural Economics*, Vol. 26, No.3.

Anderson, J. and G. Feder (2007), *Agricultural extension*, In Handbook of agricultural economics, Vol. 3 (eds. R. Evenson and P. Pingali), Elsevier Science, Amsterdam.

Black, A. (2000), "Extension theory and practice: a review", *Australian Journal of Experimental Agriculture*, Vol. 40, No. 4.

Barnes, A., J. Willock, L. Toma, C. Hall (2011), "Utilising a farmer typology to understand farmer behaviour towards water quality management: Nitrate Vulnerable Zones in Scotland", *Journal of Environmental Planning and Management*, Vol. 54, Issue 4.

Bartoli, L. and G. la Rocca (2012), "Impact of Agricultural Extension Services on Sustainable Agriculture in Italy", *Journal of External Systems*, Vol. 28, No. 1.

Birner, R., K. Davis et al. (2009), "From best practice to best fit: a framework for designing and analyzing agricultural advisory services worldwide", *Journal of Agricultural Education and Extension*, Vol. 15, No. 4.

Birner R, K. Davis et al. (2006), *From "best practice" to "best fit": a framework for designing and analyzing pluralistic agricultural advisory services worldwide,* EPTD Discussion Papers 05, International Food Policy Research Institute (IFPRI), Washington.

Blackstock, K., J. Ingram, et al. (2010), "Understanding and influencing behaviour change by farmers to improve water quality", *Science of the Total Environment*, Vol. 408.

Burton, R., C. Kuczera, and G. Schwarz (2008), "Exploring Farmers' Cultural Resistance to Voluntary Agri-environmental Schemes", *Sociologia Ruralis*, Volume 48, Issue 1.

Burton, R., J. Dwyer et al. (2006), *Influencing positive environmental behaviour among farmers and land managers – a literature review,* Countryside and Community Research Institute and Macaulay Land Use Research Institute, Aberdeen, Scotland.

Buttel, F. (2001), "Has environmental sociology arrived?", paper presented at the Annual Meeting of the American Sociological Association, Anaheim, United States.

Cameron, D. and S. Chamala (2004), "Measuring impacts of an holistic farm business management training program", *Australian Journal of Experimental Agriculture*, Vol. 44, No. 6.

Charatsary, C., A. Papadaki-Klaudianou and A. Michailidis (2011), "Farmers as consumers of agricultural education services: willingness to pay and spend time", *Journal of Agricultural Education and Extension*, Vol. 17, No. 3.

Collier, A., A. Cotterill et al. (2010), *Understanding and influencing behaviours: a review of social research, economics and policy making in Defra,* http://archive.defra.gov.uk/evidence/series/documents/understand-influence-behaviour-discuss.pdf.

Damianos, D. (2015), "An Analysis of the Performance of Farm Advisory Services as Related to Fostering Green Growth in Agriculture – The Case of Greece", Consultant Report, OECD internal document.

Dwyer, J. and M. Reed (2015), "Promoting Green Growth in England and Welsh Agriculture: Evaluating the Role of "Soft" Measures – Training, Advice and Extension", Consultant Report, OECD internal document.

Department for Environment, Food and Rural Affairs (DEFRA), (2010), *Agricultural Advisory Services Analysis*, July, http://archive.defra.gov.uk/foodfarm/landmanage/climate/documents/advisory-analysis.pdf.

DEFRA (2007), *Understanding and Influencing Positive Behaviour Change in Farmers and Land Managers*, http://webcache.googleusercontent.com/search?q=cache:nAEDBgOVW0YJ:randd.defra.gov.uk/Document.aspx%3FDocument%3DWU0104_6750_FRP.doc+&cd=1&hl=en&ct=clnk&gl=fr.

Ekboir, J. (2003), "Why Impact Analysis Should Not Be Used for Research Evaluation and What the Alternatives are", *Agricultural Systems*, Vol. 78.

Task Force on Impact Assessment and Evaluation, European Initiative for Agricultural Research for Development (EIARD) (2003), "Impact Assessment and Evaluation in Agricultural Research for Development", *Agricultural Systems*, Vol. 78.

Eurostat (2013), *Agr-environmental indicator – farmers' training and environmental farm advisory services*, http://epp.eurostat.ec.europa.eu/statistics_explained/index.php/Agri-environmental_indicator_-_farmers%E2%80%99_training_and_environmental_farm_advisory_services.

Evenson, R. (1996), "The economic contributions of agricultural extension to agricultural and rural development", in FAO (1996), *Improving Agricultural Extension: a Reference Manual*, 3rd Edition, Food and Agricultural Organisation Publications, Rome.

Faure, G., Y. Desjeux and P. Gasselin (2012), "New Challenges in Agricultural Advisory Services from a Research Perspective: A Literature Review, Synthesis and Research Agenda", *Journal of Agricultural Education and Extension*, Vol. 18, No.5.

Fazio, R., S. Farm et al. (2010), *Barriers to adoption of sustainable agriculture practices: Working Farmer and Change Agent Perspectives*, Final Report to the Southern Region Sustainable Agriculture Research and Education Program (SARE) and the Alabama Agricultural Experiment Station.

Gadrey, J. (2000), "The characterisation of goods and services: an alternative approach", *Review of Income and Wealth*, Vol. 46, No. 3.

Garforth, C, B. Angell, J. Archer and K. Green (2002), *Improving access to advice for land managers: a literature review of recent developments in extension and advisory services*, DEFRA Research Project KT0110, London.

Gowdy, J. (2008), "Behavioural economics and climate change policy", *Journal of Economic Behaviour and Organization*, Vol. 68, doi:10.1016/j.jebo.2008.06.011.

Hall, A., B. Yoganand, R. Sulaiman, S. Raina, C. Prasad, C. Naik and N. Clark (eds.), (2004), *Innovations in Innovation: Reflections on Partnership, Institutions and Learning*, International Crops Research Institute for the Semi-Arid Tropics, Patancheru, India.

Hoffmann, V., M. Gerster-Bentaya, A. Christinck, and M. Lemma (eds.), (2009), *Rural Extension. Volume 1: Basic Issues and Concepts*, Margraf Publishers, Weikersheim.

Holloway, G. and S. Ehui (2001), "Demand, Supply and Willingness-to-Pay for Extension Services in an Emerging-Market Setting", *American Journal of Agricultural Economics*, Vol. 83, No. 3.

Ingram, J., P. Gaskell, J. Mills and C. Short (2013) "Incorporating agri-environment schemes into farm development pathways: a temporal analysis of farmer motivations", *Land Use Policy*, Vol. 31, http://dx.doi.org/10.1016/j.landusepol.2012.07.007.

Kahneman, D. (2011), *Thinking, Fast and Slow*, Farrar, Straus and Giroux, New York.

King C., J. Gaffney, J.Gunton (2001), "Does participatory action learning make a difference? Perspectives of effective learning tools and indicators from the conservation cropping group in North Queensland, Australia", *Journal of Agricultural Education and Extension*, Vol. 7, No. 4.

Labarthe, P. (2005a), "Performance of Services: A Framework to Assess Farm Extension Services", Paper prepared for presentation at the 11th seminar of the European Association of Agricultural Economists (EAAE), *The Future of Rural Europe in the Global Agri-Food System*, Copenhagen, Denmark, August 24-27, 2005 http://ageconsearch.umn.edu/bitstream/24712/1/cp05la01.pdf.

Labarthe, P. (2005b), «Trajectoire d'innovation des services et inertie institutionnelle: dynamique du conseil dans trois agricultures européennes », *Géographie, Economie, Société*, Vol. 73, No. 3.

Labarthe, P. and Laurent, C. (2013), "Privatization of agricultural extension services in the EU: Towards a lack of adequate knowledge for small-scale farms?", *Food Policy*, Vol. 38, http://dx.doi.org/10.1016/j.foodpol.2012.10.005 .

Lohr, L. and T. Park (2003), "Improving extension effectiveness for organic clients: Current status and future directions", *Journal of Agricultural and Resource Economics*, Vol. 28, No. 3.

Ludwig, B. (2007), "Today is yesterday's future: globalizing in the 21st century", *Journal of International Agricultural and Extension Education*, Vol. 14, No. 3.

Marsh, S., D. Pannell and R. Lindner (2004), "Does agricultural extension pay? A case study for a new crop, lupins, in Western Australia", *Agricultural Economics*, Vol. 30, No. 1.

Marsh, S. and D. Pannell (2000), "Agricultural extension policy in Australia: the good, the bad and the misguided", *Australian Journal of Agricultural and Resource Economics*, Vol. 44, No.4.

Mills, J., P. Gaskell et al. (2013), *Farmer attitudes and evaluation of outcomes to on-farm environmental management*. Report to DEFRA. CCRI: Gloucester.

Nimmo-Bell (2009), *Evaluation of Sustainable Farming Fund Projects*, prepared for MAF Sustainable Farming Fund, Nimmo-Bell & Company Ltd., New Zealand.

OECD (2013), *Agricultural Innovation Systems: A Framework for Analysing the Role of the Government*, OECD Publishing, Paris;
DOI: http://dx.doi.org/10.1787/9789264200593-en.

OECD (2012) *Agricultural Policy Monitoring and Evaluation 2012: OECD Countries*, OECD Publishing, Paris.
DOI: http://dx.doi.org/10.1787/agr_pol-2012-en.

OECD (2009), *Methods to Monitor and Evaluate the Impacts of Agricultural Policies on Rural Development*, OECD, Paris, http://www.oecd.org/agriculture/44559121.pdf.

Pannell, D., G. Marshall, et al. (2006), "Understanding and promoting adoption of conservation practices by rural landholders", *Extension*, Vol. 46, No. 11.

PROAKIS (Prospects for Farmers' Support: Advisory Services in the European Agricultural Knowledge and Information Systems) (2014), EU funded research, 7th Framework Programme, http://www.proakis.eu/.

Prokopy, L., K. Flores, D. Baumgartner-Getz and D. Klotthor-Weinkauf (2008), "Determinants of agricultural best management practice adoption: Evidence from the literature", *Journal of Soil and Water Conservation*, September/October, Vol. 63, No. 5.

Röling, N. (1990), "The Agricultural research-technology transfer interface: a knowledge systems perspective", In *Making the Link: Agricultural Research and Technology Transfer in Developing Countries*. Westview Press, Boulder.

Rogers, E.M. (2003), *Diffusion of Innovations*, The Free Press, New York.

Ruttan, V. (1996), "Induced Innovation and Path Dependence: A Reassessment with Respect to Agricultural Development and the Environment", *Technological Forecasting and Social Change*, Vol. 53, No. 1.

Scoones, I. and J. Thompson (eds.) (2009), *Farmer first revisited: Innovation for agricultural research and development*, CTA Publishing, United Kingdom.

SOLINSA (2014), *Support of Learning and Innovation Networks for Sustainable Agriculture*, EU funded research, 7th Framework Programme, http://www.solinsa.org/.

Southerton, D., A. McMeekin and D. Evans (2011), *International Review of Behaviour Change Initiatives: Climate Change Behaviours Research Programme*, Scottish Government Social Research.

Spielman, D. (2005), *Innovation Systems Perspectives on Developing-Country Agriculture: A Critical Review*, International Food Policy Research Institute (IFPRI), ISNAR Discussion Paper 2, http://www.cgiar-ilac.org/files/Spielman_ISNAR_DP2_Innovation.pdf.

Statistics Canada (STC) (2011), *Farm Environmental Management Survey*, Ottawa, Canada.

Stubbs, M. (2010), *Technical Assistance for Agriculture Conservation*, Congressional Research Service. www.crs.gov.

Swedish Board of Agriculture (2011), *Analys av kompetensutvecklingen inom Landsbygdsprogrammet - Fördjupning av rapport 2010:30* [Analysis of advisory service within the Rural Development Programme – In depth analysis of report 2010:30] Report 2011:39.

Swedish Board of Agriculture (2010), *Kompetensutveckling inom landsbygdsprogrammet – Rapport från en statistisk undersökning genomförd våren 2010* [Advisory service within the Rural Development Programme – Report of a statistical survey in the Spring of 2010]. Report 2010:30.

Tchuisseu, R. and P. Labarthe (2015), "Investing in Knowledge to Support the Adoption of Environmentally-friendly Practices: Lessons Learned from the Canadian Growing Forward Program", Consultant Report, OECD internal document.

University of Gloucestershire, Countryside and Community Research Institute, and Macaulay Land Use Research Institute, Aberdeen (CCRI/MLURI) (2006), *Influencing positive environmental behaviour: behaviour among farmers and landowners – a literature review*, Final Report Annexes 1 and 2, Gloucestershire, UK.

Waddington, H., B. Snilstveit, H. White and J. Anderson (2010), *The Impact of Agricultural Extension Services,* 3ie Synthetic Reviews SR00 9 Protocol, International Initiative for Impact Evaluation, www.3ieimpact.org/media/filer/2012/05/07/009%20Protocol.pdf.

Chapter 5

Investing in knowledge to support environmentally-friendly practices: Lessons learned from selected case studies

This chapter summarises the key findings and lessons learned from five case studies: the Australian Primary Industries Research, Development and Extension System; The Canadian Growing Forward Program; "Soft" measures in England and Wales; the Farm Advisory Services in Greece; and the Sustainable Farming Fund in New Zealand.

Five case studies were undertaken to briefly illustrate different approaches and best practices in advice offered to farmers and the policy initiatives taken to encourage green growth. The Australian case study outlines the evolution, design features and performance of the country's research, development and extension system and the improved industry competitiveness and sustainability that this has contributed to. The study on England and Wales provides insights into the operation, strengths and weaknesses of their agricultural knowledge system in order to assess the value of the advice offered, as well as of other "soft" policy mechanisms that provide or stimulate knowledge exchange and skills acquisition to support the transition to sustainable agriculture. The Greek case study analyses the performance of the Farm Advisory System (FAS), set up in 2007 in compliance with the 2003 EU CAP reforms and designed to be independent of the previous extension system. It was found, however, that best-fit policy practices by the FAS are proposed in a setting of institutional inaction, inadequate transfer of knowledge, and unused agricultural potential. The New Zealand case study examines the experience of the Sustainable Farming Fund, a main source of funding for agri-environmental innovation and research. This case study highlights the importance of community-led research projects in contributing economic, environmental and social benefits to a country's primary industries and rural communities. Finally, the Canadian case study explores the issue of investment in knowledge to support the adoption of environmentally-friendly practices by analysing the Canadian Growing Forward Program in three provinces: Ontario, British Colombia and Quebec. The methodology for all case studies combines a literature review, interviews and policy document analysis.

The Australian primary industries research, development and extension system

Key messages

- The research, development and extension (RD&E) system of primary industries is an important driver of innovation and improved competitiveness and sustainability of the Australian agricultural sector. Its strength lies in the partnerships generated between the rural Research and Development Corporations (RDC), government, farmers and private sector parties, all of whom contribute to identifying and prioritising issues requiring RD&E.
- The RDC model ensures that producers who benefit from research also contribute to its costs. Periodical evaluations of the performance and impact of the Australian RD&E system suggest that the returns from investments, gains in productivity and improvements of the environment are all significant, although difficult to quantify.
- Key challenges pertain to maximising the payoffs to public investments, while minimising transaction costs across the multiple R&D and extension providers and jurisdictions that comprise the Australian RD&E system. This includes the need to improve the balance between different types of public investment (e.g. R&D versus extension; focus on food quality and food safety management across value chains versus a natural resource management focus; cross-cutting versus commodity-specific R&D).

Description

Under national and state government arrangements, Australia has developed a unique system for primary industries RD&E, which involves a number of actors working together. A range of programmes are in place across several government departments which provide funding for rural RD&E. A significant proportion of Australian government financing on rural RD&E is conducted through rural RDC, within the Agriculture, Fisheries and Forestry portfolio.

The RD&E system functions under Australian government legislation, which facilitates the collection of levies from industries. The collected levies are matched by the Australian government. These funds are managed by the rural RDC which co-invest with RD&E service providers, such as state and territory government agencies, universities and the private sector. In recent times, this system has evolved under a variety of influences to meet the changing needs of its stakeholders.

The extension model post-1950 in Australia was more linear – that is, scientists producing research and delivering it to farmers – and (local or state) government agencies dominated service provision. Over time, however, government investments in extension services have gradually diminished, while the private sector played a much more prominent role in delivering a wide range of extension services. Industries and some jurisdictions have varied in the rate of increase of the private sector involvement, leading to different approaches to extension across the country. For example, the grains sector relies almost entirely on the private sector for extension services via private consultants or private farmer group structures, while governments tend now to focus their services on providing specialist services (e.g. biometrics) and extension in biosecurity and public good areas, such as natural resource management.

The originality of the Australian RD&E system is the rural RDC model of co-financing of rural R&D activities, established in 1989. The rural RDC model is a public-private partnership between the Australian government and private industry, whereby the RDCs procure rural R&D, using funds collected from processors and paid either by processors or primary producers – depending on the industry – as well as funding provided by the government, with maximum matching contribution per year of 0.5% of an industry's gross value of production. There are 15 RDCs that cover virtually all agricultural industries, as well as fisheries and forest and wood products (Grant, 2012). The main advantages of the RDC co-investment model include the following:

- Helps to ensure that all producers who benefit from research contribute to its cost, and addresses free-rider problems that could lead to under-investment in R&D.
- Helps to elicit additional, socially valuable R&D, including situations in which the benefits are either spread thinly across a wide range of industries, or mainly accrue to the wider community.
- Helps to ensure that public money is not spent on research of little practical value.
- Facilitates greater and faster uptake of research outputs (OECD, 2015a).

While RDCs are the cornerstone of the national RD&E system, it was not until the establishment of the National Primary Industries Research, Development and Extension Framework (NPIRDEF) that a genuinely integrated, national primary industries RD&E system existed. The NPIRDEF, which was a product of a 2009 agreement among all major stockholders in Australia's Agricultural Knowledge System (i.e. the federal and state governments, RDCs, the Commonwealth Scientific and Industrial Research Organisation and the Australian Council of Deans of Agricultural Universities) aims to encourage greater collaboration and promote continuous improvement in the investment of RD&E resources nationally (Framework, 2014). Thus far, the NPIRDEF comprises 14 sector strategies (being one for each RDC, such as dairy, grains, sugar, cotton, wool) and eight cross-sector strategies (animal welfare, plant biosecurity, animal biosecurity, food and nutrition, water use, climate change, biofuels and bioenergy, and soils).

A number of improvements in the national RD&E system became evident as the sector and cross-sector strategies were developed and implemented including: agreed national strategies; broader, more inclusive governance arrangements involving a larger number of relevant parties; a more strategic engagement between industries and governments; and more stable funding arrangements allowing for longer-term investments, better resource planning and improved career security for staff; strengthened national research capabilities to better address sector and cross-sector issues; and focused RD&E resources ensuring they are used more effectively, efficiently and collaboratively, thereby reducing fragmentation and duplication in primary industries RD&E (OECD, 2015a).

Notwithstanding this progress, there are still a number of challenges to address: the effort of managing strategy implementation is challenging given the wide range of partners involved; cross-sector strategies are inherently more complex; the perceived reduction of public funds, while objectives are broadening; and the need to improve the balance between different types of public investment (e.g. R&D versus extension, cross-cutting versus commodity-specific R&D, etc.) (Grant, 2012).

Evaluation

Evaluation of the performance and impact of the Australian RD&E system has been conducted periodically at a range of levels, including:

- RD&E system evaluations (e.g. Productivity Commission, 2011; Allen, 2012).
- Portfolio evaluations (covering collective RDC or cross-sector investments) (e.g. Acil-Tasman, 2008; 2010).
- Industry evaluations (covering particular industry sectors).
- Programme evaluations (covering medium- to long-term programmes, often comprising multiple projects).
- Project evaluations (covering individual projects, not included here).

There have been a large number evaluations of Australian primary industries RD&E performance which share similar findings overall. While the estimated economic returns vary between projects, the overall returns are considerable, with benefit-cost ratios supporting the investments. Evaluations of social and environmental returns are fewer, more qualitative and commonly associated with economic returns.

The portfolio/industry/programme-level case studies show that substantial productivity, environmental and social benefits have been generated for industry and the public, usually simultaneously. However, it is not possible to quantify certain benefits, especially in environmental and social areas. Conservative estimates of benefits typically outweigh costs by greater than 3:1. For example, case studies report the doubling of milk production over 30 years; the revitalisation of the lamb industry; improved chemical management in the cotton industry; more sustainable grain production; a wide range of improvements in natural resource management; and important social benefits. These benefits are typically delivered simultaneously.

The RDC model is regularly reviewed. The Productivity Commission, which is the Australian government's independent research and advisory body on a range of economic, social and environmental issues, conducted an inquiry into the RDC model in 2010 (Productivity Commission, 2011). It estimated the total recurrent costs at approximately AUD 1.5 billion annually and also explored the possibility of reducing the number of RDCs as well as the potential savings associated with consolidating "back of office" operating overheads.

This inquiry supported much of the RDC model. No strong argument to reduce the number of RDCs was found, but it was recommended that attention be given to efficiencies in overheads and administrative functions. The level of "additionality" attached to the government investment and the distribution of benefits relative to the levels of funding were challenged. The report inquiry argued that the benefits flowing to industry were significant and sufficient to attract more industry investment. The returns to the public were also seen as modest relative to the shares of funding (governments 76%, industry 24%), which highlights the risk of a certain degree of "crowding out" of private sector investment.

Two detailed portfolio evaluations were undertaken to measure the economic, environmental and social returns of R&D investments by the RDCs, using cost-benefit methodologies across selected projects (Acil-Tasman 2008; 2010). These are reported to be the largest evaluations of rural R&D undertaken in Australia up to date. The 2008 review examined 36 "highly successful" projects that could demonstrate significant evidence of delivery and 32 "randomly selected" projects, from a pool of 600. The randomly selected projects were expected to provide an indication of the average returns from investments and the highly successful projects an indication of the potential returns. They were selected from across the full range of RDC fields of interest. A counterfactual was also derived to assess the likely outcome in the absence of the interventions. Economic, environmental and social impacts are considered, quantified where possible, or assessed qualitatively where this was not possible.

A similar evaluation of RDCs was undertaken in 2009 with 59 randomly selected programmes representing AUD 676 million investments in forestry, meat, fodder crops, soil biology, education and fisheries management. The results show benefits outweighing costs 2.4:1 after five years; 5.6:1 after ten years; and 10.5:1 after 25 years, very similar results to the 2008 evaluation. The value of using the series of time horizons is to show the accumulation of benefits over time, with 60% of projects being net present value positive by year 5 and 76% by year 10.

Overall, the nature of RD&E makes estimating returns from investments more difficult and very challenging. RD&E in primary industries routinely takes two or three decades to achieve widespread adoption (Alston et al., 2011). These lengthy periods allow for many interventions, both positive and negative, often by many actors, some which are uncontainable (e.g. weather). Separating these effects can be difficult (if not impossible). Moreover, adequate data is not available to assess performance against implementation or non-economic social outcomes (Cuevas-Cubria et al., 2012). Nevertheless, although the benefit-cost assessments may over-estimate the benefits to some degree, there is substantial evidence that the economic returns from RD&E in primary industries are large, relative to the investment.

Lessons learned and policy recommendations

Time and resources

Evaluation in this field is more difficult, due to timeframe and attribution complexities and the lack of a quantifiable value for some important benefits (e.g. some environmental and social outcomes). Notwithstanding this, substantial evidence is provided from portfolio, industry and programme impact assessments, that the returns to the economy, environment and society are large, relative to the investment with benefit-cost ratios routinely exceeding 3:1. The change can take several years to build sufficient support and change momentum. Adequate levels of appropriate resources are needed to support model development, the change process and on-going implementation.

Industry and government funding

Arrangements to bring industry and government resources together have a number of important benefits, including: increasing the total available resources; expanding the role of industry in strategy development and priority setting, and thereby improving grower engagement and adoption; enabling industry to contribute to industry benefit outcomes; and allowing broader public good benefits, such as natural resource management, to be addressed in combination with industry productivity benefits. It is very important that all the relevant parties and resources necessary to run the model are captured under these arrangements or there is a risk of losing crucial capability.

Institutional arrangements

Entities with responsibility for each of the sector or cross-sector area (like the RDCs) are very important in segregating the work, focusing the strategies and effort, as well as ensuring comprehensive coverage and adequate attention to all aspects of the overall objectives. An agreement that captures the clear purpose, intention, objectives and processes of the parties outlines the rationale for collective action and aligns effort.

Agreed national strategies align and consolidate effort and generate significant efficiency gains. The development of each strategy has involved assessing future capability needs against that currently available. In most cases remedial action was initiated to develop and secure the capability needed. Capability no longer required could then be redirected or rationalised to provide for more relevant resources.

A joint management function (including the major investors) is required to provide ongoing system oversight to assist strategy development, system improvements and to report to stakeholders. Good systems and processes are important to ensure that resources are used efficiently and progress is made.

Impact on productivity, the environment and socially

The returns from investments overall are very significant and provide gains in productivity, the environment and socially (often simultaneously) in priority areas targeted by governments and industry. The returns are sometimes hard to quantify accurately; however, even conservative assessments show that they significantly outweigh the costs and often justify increased investment.

Strategic approach

A coherent approach which combines a strategic (top-down) approach with input from industry and service providers (bottom-up) appears to be more sustainable and provides for better performance than top-down approaches alone. The strength of the Australian RD&E system lies in the partnerships that are generated between industry, government, farmers and private-sector parties, all of which contribute to the identification and prioritisation of issues requiring RD&E.

Private sector role

The role of the private sector is increasing in RD&E and should be encouraged while ensuring that "crowding out" is minimised, knowledge-sharing is maximised, relationships are co-operative and roles are complementary. Governments today rely on this important private sector contribution and it is noteworthy that some important innovations will not reach the market without it.

Evaluation methodologies

Evaluations add most value when qualitative and quantitative analyses are combined to provide a richer appreciation of what is achieved. Sole reliance on individual metrics risks under-estimating the value and can drive counter-productive goal displacement. Attribution to a particular investment is also often fraught. Frequently, R&D from other countries contributes to the outcome. Perhaps what matters more is that these investments add considerable benefit over the long term and that a team effort is required to deliver them.

Improving cost-effectiveness

While the return on investment from the Australian RD&E system is good, it is important to continue paying attention to running costs.

The Canadian Growing Forward Program

Key messages

- In the context of decentralisation, the level of support to training and advice facilitating the adoption of agri-environmental measures depends on the setting of national priorities and of specific local political agenda in each province. Overall, the large-scale farmers tend to participate more than small-scale ones, and their operations are expected to have a higher impact regarding environmental objectives.
- The effectiveness of the policy in facilitating knowledge flows within the AKIS depends on the relations between farmer unions and the administration, and on the degree of cohesion within farmer organisations. These relations seem to be strongest in Quebec. In all three provinces examined (British Columbia, Ontario and Quebec), farm advisors are well-trained.

Description

In Canada, agriculture is a shared jurisdiction under the country's constitution. The general policy framework is governed by joint Federal, Provincial and Territorial (FPT) agreements. The current five-year policy framework for food and agriculture is Growing Forward II, which replaced the 2003-08 Growing Forward policy framework program (AAFC, 2013). Both policy frameworks emphasise the importance of innovation, competitiveness, market access, sustainability and adaptability (for a review of the role of the government in fostering innovation in the Canadian agri-food sector see, OECD, 2015b).

These agricultural policy frameworks also include funding for the development of farmers' skills and knowledge. Under the Agricultural Policy Framework (2003-08), some programmes (or programme components) were delivered directly by the federal government, while Growing Forward 2 (2008-13) transferred the delivery of these programmes to the provinces and territories, allowing for more flexibility and making the programmes more responsive and adaptable to actual needs.

This case study examines the investment in knowledge to support the adoption of environmentally-friendly practices by analysing the specific case of the Canadian Growing Forward Program (GFP) in three Canadian provinces – British Columbia, Ontario and Quebec. The study aims to contribute to the debate about the relative effectiveness of supply-driven versus demand-driven policies regarding farm advice or training related to agri-environmental measures. A key feature of the analysis is that it takes into account the fact that these services are embedded in AKIS (OECD, 2012).

The GFP case study was chosen for three reasons: Firstly, the GFP deals with the specific Canadian environment and its current challenges in the realm of sustainable agriculture. Secondly, agri-environmental measures, including training for advisory services, are key elements of the GFP. Thirdly, the Canadian context offers great potential to support a discussion of the evaluation of agri-environmental policies. To achieve agricultural sustainability, many regions of Canada have implemented programmes based on a combination of the sustainable management of resources and monitoring procedures – which can serve as examples of the concerted management approach (Summers et al., 2008; Eilers et al., 2010). Finally, Canada has implemented a precise system for monitoring the impacts of agri-environmental policies, including training and advice.

It was decided to analyse this programme in three provinces: Ontario, British Columbia and Quebec. First, Ontario draws our attention because it was in Ontario that the Environmental Farm Plan (EFP) was inaugurated in 1993, as a pilot project, at the request of the farming community, and with an incentive to protect the environment (Summers et al., 2008). In fact, the Ontario EFP has played an important role in the definition of Canadian agricultural policy, and was defined by Agriculture and Agri-Food Canada (AAFC) in 2003 as one of its crucial environmental components (Fitzgibbon et al., 2004). Thirty-five percent of Ontario producers had completed an EFP in 2006, and another 6% had EFP under development (Statistics Canada, 2007). In Ontario, the GFP, started in 2008, is the continuation of the EFP. The programme provides financial payments to producers, processors, organisations and other actors in the agriculture and agri-food industry with a special focus on innovation, competitiveness and market development.

Secondly, attention is focused on British Columbia because the EFP was launched in the province in 2003, with the decision of AAFC to adopt EFP as a key component of its environmental framework policy. According to Statistics Canada (2007), in 2006, 11% of producers in British Columbia had completed EFPs and another 9% had plans under development. The GFP in British Columbia is based on a demand-driven conception for the provision of farm advisory services, through EFP. Upon request, farmers can receive funding for implementation of environmental risk reduction projects.

Thirdly, the Quebec case was also selected for two reasons. First, Quebec is Canada's only predominately French-speaking province. Second, the conception of the support to advisory services is more supply-driven, with the funding of advisory clubs – an infrastructure of AKIS that has existed in Quebec since 1993, with agricultural advisors in the advisory clubs producing agri-environmental farm plans. The other reason behind the choice of Quebec is that, according to Statistics Canada (2007), Quebec's producers had the highest rate of adoption of EFPs in Canada in 2006, with 73% having completed an EFP, and another 4% having plans under development.

Quebec's GFP includes a substantial investment aimed at supporting growth and at ensuring the sustainability and prosperity of Quebec's agriculture and agri-food industries. Support is given to farmers mainly through advisory clubs. The implementation of this programme is achieved through: a network of advisory clubs under the *Prime Vert* programme[1]; Quebec's plant health agriculture strategy (reduction of pesticide use and promotion of integrated pest management); research; knowledge transfer and cross compliance.

In total, a comparison of the implementation of EFP and GFP in the three provinces would make it possible to put into perspective those programmes that share the same orientation and objectives, but have different intervention axes are that may lead to different impacts of the farm advisory services. The conception of the intervention is different: it is, for instance, more demand-driven in British Columbia, with the support focused towards farmers' individual demands, whereas in Quebec it is based more on the support of existing infrastructures of AKIS, such as farmers' advisory clubs.

In this perspective, a simplified conceptual analysis framework, differentiating six intervention axes has been developed to assess public programmes supporting farm advice. The methodology used combines a systematic literature review, interviews and policy document analysis. The direct data collection method was limited to semi-structured interviews (face-to-face and by telephone) conducted in February 2014.

Evaluation

The framework is based on three scientific articles that describe the i) "fit for purpose" criteria of advisory services (Prager et al., 2014), ii) a framework for designing and analysing pluralistic agricultural advisory services from "best practice" to "best fit" (Birner et al., 2006), and iii) the economy of services and the public policies of agricultural advisory services (Laurent and Labarthe, 2011).

Thus, the proposed framework for analysing agricultural services policies is conceptualised, based on the reliability of the articles found in the literature, to provide an overview on six intervention axis. Two of these axes describe the context in which the farm advisory programme is implemented, that is: the policy development strategy (i.e. the political system that can be federal, such as in Canada, the United States and Germany or more centralised as in France) (intervention axis 1) and the farm structures and environment (intervention axis 2).

Three other axes describe the potential inter-linkages of the different elements of the farm advisory system by: the configuration and the capacity of agricultural service provides and other actors of the AKIS system (intervention axis 3); the overall approach used by agricultural advisors (i.e. supply- or demand-driven services (intervention axis 4); and the methods used by advisors to update their knowledge (intervention axis 5). The last axis focuses on the evaluation procedures in place for the programme under study (intervention axis 6).

Table 5.1 indicates for each intervention whether the empirical material gathered enabled to assess the effects of agricultural advisory services on agri-environmental practices in Ontario, British Columbia and Quebec.

Table 5.1. Evaluation of the criteria for assessing the impacts of advisory services on agri-environmental practices

Intervention axis	Assessment
1. The policy development strategy	+
2. The production systems and socio-economic conditions in which the farmers operate	+
3. The configuration and the capacity of agricultural services providers and other actors of the AKIS	+/-
4. The approach used by agricultural advisors: demand-driven versus supply-driven	+/-
5. The methods used by advisors to update their knowledge	+/-
6. The procedures in place for the impact assessment of agricultural advisory services	-

Note: + indicates that the objective has been met; - indicates that the objective has not been met; +/- means inconclusive.

Lessons learned and policy recommendations

- In a context of decentralisation, the level of support to training and advice facilitating the adoption of agri-environmental measures depends on the setting of national priorities and of the specific local political agenda in each province. Overall, large-scale farmers tend to participate more than small-scale farmers, and it is expected this will have a higher impact regarding environmental objectives. Such a view is a matter of debate among Canadian scholars.
- The effectiveness of the policy in facilitating knowledge flows within the AKIS still depends on the relations between farmers' unions and the administrations, and on the degree of cohesion within farmers' organisations. These relations seem to be stronger in Quebec, which has one strong farmers' organisation, than in Ontario or British Columbia. In the three contexts, industries providing inputs to farmers have a direct and indirect influence on other AKIS stakeholders.
- There seems to be a trend towards more demand-driven conceptions of advisory and training services, where the role of administration would be focused on guaranteeing the quality of advice, providing financial support to farmers' demands for services and monitoring the delivery and funding system.
- There many well-trained farm advisors in all three provinces who provide advice on agri-environmental issues. Their sources of knowledge comprise of public authorities, public research centres, non-governmental research centres, private consultancies, private companies, farming publications and social media. But there is a lack of information regarding how this knowledge is updated.
- In general, the evaluation of the programme is based on the satisfaction of the participants with the programme and on the percentage of adoption of agri-environmental measures. But there is no evaluation of the impact of advisory services on changes in practices or on the environmental performance of farms. This case is no exception, and confirms a gap found in scientific literature about the effectiveness of farm advisory services for integrating environmental issues.

In interpreting the results of this case study, it is important to take into consideration its strong limitations. The sample depth and breadth cannot be considered as representative. Firstly, the study considered only three provinces. Secondly, the interviews were completed exclusively with experts from universities. It is recommended that future studies should include in the sample all the of the AKIS agents and should provide in-depth analysis of the interactions and influence among them. There is currently also significant knowledge gap in the scientific literature regarding the evaluation of the impact of the educational dimension of agri-environmental policies (farm advice and training), which merits further research.

"Soft" measures in England and Wales

Key messages

- The AKIS in the United Kingdom is diverse with respect to strategies (i.e. determining what types of agricultural knowledge are needed and for what purpose), the type of agents and methods deployed to deliver advice, information and training.
- To date, no studies have provided sufficient evidence on the costs and benefits of these measures that would enable a full assessment of cost-effectiveness. Available evaluations focus largely on "snapshot" evidence and often based on a low number of participants in interviews and surveys.
- Qualitative evidence suggests that soft measures can be a vital element in supporting the transition towards sustainable agriculture. They appear to be most effective when they focus on promoting a common set of practical approaches to farmers.

- Policy makers should pay increased attention to the barriers which prevent or discourage farmers from taking the advice, training and information offered by private and public services, and should design targeted approaches to overcome these barriers, learning from existing best practices and being sensitive to the context(s) within which they operate.

Description

The main objective of this case study is to provide insights into the operation, strengths and weaknesses of the agricultural knowledge system in England and Wales. It seeks to assess the value of "soft" measures, such as advice, training and extension policy mechanisms, as a means of supporting the transition towards sustainable agriculture. These policy approaches work on the basis of providing or stimulating knowledge exchange and skills acquisition as a trigger to changes in practice, in contrast to those which contract directly to support changes in practice with payment, via mechanisms such as agri-environmental schemes (AES) or farm investment grants; or those which seek to achieve change through prohibition, conditions or disincentive such as regulating the use of certain agrochemicals, or applying eco-taxes.

In England, the state-funded Agricultural Knowledge System (AKS) was dismantled in the late 1980s. The state-funded Agricultural Development and Advisory Service (ADAS) was largely privatised and state-funded research went through a period of review and consolidation. The retreat of government from the management of agricultural research and extension resulted in a diversification of the sources of agricultural research and extension, and opened new opportunities for the private sector.

Perhaps the most well-known initiatives in England since 2000 have centred on addressing diffuse agricultural pollution, in the specific context of targets set within the EU Water Framework Directive and an expanded coverage of Nitrate Vulnerable Zones under the EU Nitrates Directive following policy reviews, in 2002. The Catchment-Sensitive Farming Demonstration Initiative represents one of the longest-established such programmes, where advisers working within priority catchments are tasked with facilitating improved understanding and standards of practice on farms within these areas.

In 2013, Defra published a review of policy on advice and advisory services which re-emphasised that the government seeks to avoid the "crowding out" of private sector advisors by making its documentation "open source" and of only providing public-funded advice in instances of clear market failure (Defra, 2013). Defra also signalled a 25% reduction in its expenditure on farm advice (down to GPB 15 million per annum), and primarily through enhanced on-line provision of information. The department is emphasising the importance of partnership with private and third sector bodies in order to ensure effective knowledge diffusion within agriculture.

Its aim is that advice that meets the public needs will be delivered in a flexible way through a variety of channels with "realistic" expectations regarding participation, and that its targets are based on shared goals and simple tasks. Regulatory advice remains outside the scope of this flexible approach, but Defra suggests it will be delivered in a more integrated manner, often through partners in the industry. For the farmer, Defra suggests this will mean:

- "Government-funded advice is more widely accessible.
- More advice delivered by professionals and trusted organisations that understand local issues and concerns.
- Clear and focussed messages that are easier to implement on farm.
- Access to better local knowledge exchange and networks where farmers share ideas and learn about best practice in a practical setting.
- The individual farmer will be more empowered to act through industry-led initiatives." (Defra, 2013).

Wales, since devolution in the 1990s, has established its own distinctive approach to agriculture and farm policy. With relatively small, family farms and a predominance of livestock production, the provision of many kinds of advice and training to Welsh farmers has been seen by the Welsh government as a legitimate purpose of policy. Although the ADAS in Wales went through the same privatisation process as occurred in England, successive administrations have committed to supporting farm advisory services and initiatives in a more comprehensive manner than now prevails in England.

Of particular note in Wales was the setting up and operation of extension or coaching based upon the development of learning processes through farmer groups (Agriscôp), established to undertake collective learning as a means of enhancing business performance (Pearce and Willliams, 2010). In addition, resources were devoted to ensuring that farmers joining agri-environment schemes in Wales were obliged to undertake some basic training in environmental management, funded via rural development and cohesion policies. Finally, the Welsh government has shown long-term commitment to funding specific advisory services and/or methods (e.g. Farming Connect) that have now been in place for several decades, rather than opting to encourage competition and a wide variety of providers, as happens in England.

Evaluation

As discussed earlier, a common gap in most research on farmer advice, information and training is that quantitative methods are seldom deployed to measure impacts, not least because of the complexity – with both temporal and conceptual challenges – of so doing. It is therefore not possible to do simple cost-benefit analyses to examine the cost-effectiveness of this type of measure. Instead, a range of both quantified and qualitative data is discussed and brought together in an attempt to assess some of the main constituents, and the potential scale, of net costs and benefits.

More specifically, the analysis is based on a survey of the available qualitative evidence from documents and selected interviews concerning the impacts of these measures in promoting sustainable agriculture in England and Wales. Direct evaluation experience from Countryside and Community Research Institute (CCRI) studies has been complemented with a review of both academic and "grey" literature published by government departments, agencies, NGOs and the commercial sector, on the general topics of knowledge exchange and mechanisms to promote sustainable farming in England and Wales. Moreover, selected interviews with key informants were held to collect their experience with the provision of agricultural knowledge and extension. When combined with the findings from the comparative analysis of evidence of impacts, it is possible to make some general, albeit tentative, judgments concerning the relative effectiveness of soft measures as a mechanism to encourage sustainable agriculture, in England and Wales.

Comparative evaluation

Taking stock of the findings of the different qualitative evaluations, it seems clear that advice, training and information have been judged in nearly all the cases reviewed to have demonstrated significant benefit as mechanisms to assist or encourage transformation towards sustainable agriculture. This impact is felt at farm level, across local territories and within food supply chains.

Funding advice, training or information may not immediately result in farm changes on the ground, in terms of practices or systems. It is particularly difficult to assess net impact in cases where studies are tasked only with recording to what extent a project or initiative has met its objectives because very few such evaluations are required to consider a counterfactual.

However, several studies did ask participants to reflect on the degree of additionality involved therein. Where it was seen as significant, additionality appears stronger in approaches involving a knowledge-exchange in which farmers are directly engaged in learning processes alongside people with other kinds of expertise, rather than in more formal, uni-directional knowledge-transfer provision. Key to this observation is the fact that a learning process, in contrast to a one-off training or information event (even a single farm advice visit, in isolation), provides much more scope for the farmer to absorb, test and re-create the main messages involved in the knowledge exchange process – this can be key to what psychologists term "central route processing" of new information, increasing its likelihood of direct impact in practice.

The neglect of the learning process within the scheme and programme evaluations is considered to be a main barrier to achieving any robust assessment of the cost-effectiveness of advisory, training and extension measures. Appropriate data is not being gathered in the majority of cases and this may be related to a lack of demand for such information from the commissioning sponsors within government departments and agencies.

Gauging the quality of the evidence base represented here, it is clear that some significant gaps exist. Some are due to the fact that most evaluations have focused on individual projects or initiatives with relatively short life spans. This leads to an emphasis in the studies on using qualitative research to gather insights into farmer- or advisor-reported attitudinal and/or behavioural change, or attempting the measurement of proxy indicators of impact. The project-driven nature of the existing research tends to focus on measurements deemed most relevant to the intervention (i.e. project goals and targets), rather than giving any opportunity to consider wider outputs or more fundamental impacts (Reed and Courtney, 2013). As demonstrated in the discussion below, there is often a gap between the intended impacts of soft measures (e.g. better habitat) and what appear to be the actual outcomes of these approaches (e.g. improved confidence to act and to learn, stronger social networks). It therefore seems highly likely that, even where soft measures have been evaluated, only a proportion of their impact has been captured.

Cost effectiveness

At the macro level, data from the Agricultural Industries Confederation suggest that the input supply industries in the United Kingdom with trade valued at GBP 6.5 billion spend approximately GPB 200 million a year on advisors and representatives, investing in the region of GPB 40 million in near-market R&D in the same period. They point to "risks of up to GPB 1 000 per hectare being dependant on farmers having the correct agronomic information" (referring to the value of lost output, if inputs are inappropriately or under-used), indicating the importance of professional and research-informed advice (Gibbs, 2013). However, the same sources are not able to provide robust data on the impact of the advice provided, although it would appear that the industry itself judges the benefits of such investment to be outweighed by the costs, using illustrative figures such as these to underpin that decision.

At the other extreme, the CCRI evaluation of the non-profit networking initiative Linking Environment and Farming (LEAF), which had a very limited membership, illustrative sample size of 10 farms, suggests considerable benefits from the advice, information and networking offered, among the interviewed farmers. As part of the reported benefits were savings on fertiliser use ranging between GPB 2 500 and GPB 10 000 per farm, per year; and livestock producers reported cost savings of around 10% due to improved animal health. Others reported that the process of joining LEAF had increased the speed of integrated farming systems (IFS) adoption on their farm, so realising savings on input costs earlier than would have occurred otherwise. This was claimed to have involved a GPB 4 000 per year saving on one farm (Mills et al., 2010).The greatest barrier to achieving any robust assessment of the cost-effectiveness of "soft" measures is undoubtedly the neglect of this issue within the process of scheme and programme evaluations. Appropriate data on costs, results and outcomes (either changes in perceptions, understanding and intentions to act, or actual changes in practice) is simply not being gathered in the majority of cases. This may be related to a lack of demand for such information from the commissioning sponsors within government departments and agencies. The example of the work being done now in the evaluation of Agriscôp shows that this kind of evaluation should indeed be possible.

Lessons learned and policy recommendations

The survey of available studies suggests that there is good evidence from policy and private-sector initiatives promoting soft measures in England and Wales that these measures can be a vital element in supporting the transition towards sustainable agriculture, and that there is a clear case for a public role in these approaches based largely upon arguments of market imperfection relating to small businesses (e.g. incomplete information), as much as market failure relating to public goods.

In sum, the qualitative assessment reveals the many and varied ways in which advice, training and information are being used to promote this goal, within public, private and third sectors. It also indicates that (to an extent) the varied impacts of these tools have been recognised and documented, although there are problems and challenges in achieving any meaningful valuation or quantification of impacts, such as would ideally be needed in order to assess and compare their value in a consistent way.

However, it is also clear that no studies to date have provided sufficient information on the costs and benefits of these measures to make a full assessment of cost-effectiveness possible. This would be needed to enable some element of benchmarking, or comparison against the cost-effectiveness of other approaches including agri-environmental payments or Payments for Ecosystem Services; or funding investments to facilitate more sustainable practices.

Nevertheless, the findings reviewed and analysed in this case study do enable us to bring together several conclusions and a series of common messages in the evaluation studies, concerning how best to design and deliver effective "soft measure" approaches in policy can be identified. Table 5.2 summarises these findings.

Table 5.2. Summary of good practice in "soft measure" design and delivery

At the farm level	The community level	Project level
Advisors should be local, expert, and associated with a well-established and trusted institution.	Farming is connected to the local community; Initiatives that explore and demonstrate connections promote this benefit	There is a need to understand that incentives only become sustainable when coupled with learning
Sustainability advice should still be orientated towards building the profitability of the farm business.	Skill facilitation is the core of delivering a successful project: building capacity, confidence and reflexive capability among farmers.	Understand the needs of localities by involving local people is especially valuable
Profitability is a key way of discussing continuation of the farm - the central goal of much family farming.	Positive changes to the self-identity of farmers can be achieved in community development settings, which engender trust and encourage farmers to feel that they can do things that will really make a difference.	There is a need to build answers to environmental challenges from locally-appropriate/embedded solutions
Solutions should be produced with farmers rather than for them.	Ditto for the local community – involving them enlarges the scope of possible solutions	Clear and unambiguous communication is vital to combat "hearsay"
Problem solving should build on farmers' own knowledge and show pathways to using and improving upon that.	Knowledge from other community members can provide insights into new potential business-environment linkages on farm	Monitor during the process to understand progress, and react during the programme
Advice delivered by a small core team is trusted	Expert facilitation may be key to build effective links between farm decision-making and community goals	Spatial Integration of multiple agency goals is key – at present, many agencies seek different actions on the same areas of land, creating confusion at farm level

There is a limited evidence base in the body of evaluation studies conducted to date, which is largely focused on "snapshots"; and often based on low numbers of participants in interviews and surveys. This "deep but narrow" style of evaluation studies has however proved useful in helping industry and policy makers to understand the causal linkages between soft measures and their impacts, and to focus on how to encourage participation and communicate effectively with farm businesses.

As recognised in Defra's (2013) policy review and indicated by the figures provided by the Association of Independent Crop Consultants (AICC) in their own report, it seems clear that the role of professional advisors operating in the commercial sphere can be key to improving farm businesses through the provision of advice that maximises market performance. The recent creation by the AICC of a Register of Feed Advisors[2] extends the industry's investment in Training and Continual Professional Development (CPD) and professional standards more deeply into the livestock sector, helping to counterbalance what has traditionally been much stronger provision for cropping farms. Thus a tactic of extending the reach of professional advice into issues of sustainability, as recognised by the input industry and Defra, could be useful for policy makers.

However, commercial sensitive issues currently limit the extent to which government goals can be robustly pursued via retailer/food processor protocols (MacDonald et al., 2006); and it remains the case that the less than comprehensive nature of penetration of commercial advice across the sector, noted earlier, will limit its ability to deliver policy goals for sustainable farming.

From the evidence discussed in the case study (Dwyer and Reed, 2015), it can be concluded that significantly improved environmental results might be achieved through improving the training and advice that accompanies agri-environmental schemes, a conclusion echoed in the "making environmental schemes more effective review process" undertaken by Defra agencies, in recent years (Natural England, 2014). Training to address confidence about specific activities helps to address broader questions of participation in schemes and initiatives, even including those that were originally designed to be adopted widely without specialist advisory support.

There are considerable gains to be made in greater uptake and more effective use of existing technologies for sustainable agriculture, by including more farmers in the circles of self-help or government-supported advice and training that have already been developed, in a variety of contexts. This will help to increase the returns from those technologies.

The focus of evaluation studies on farmer participation has perhaps tended to obscure the importance of the wider context – i.e. the self-perception of the farmer in terms of their community identity, personal sense of efficacy (the notion that they can make a real difference to the environment through their own actions) and ease of access to relevant and directly applicable information, in these initiatives – in achieving positive change. The evidence examined suggests that initiatives enabling ongoing learning, with group discussions and processes with peer-support, can be more effective than conventional one-to-one, unidirectional knowledge transfer provision.

The limitations of the existing research evidence indicate that the effectiveness of different forms of training, advice and extension are not yet comprehensively understood. Public and private investment decisions in that domain are therefore being made upon a narrow evidence base, suggesting that they may not be providing optimal returns.

England and Wales provide contrasting approaches to how advice and training are delivered to farmers through public policy. Whilst public-funded provision in Wales is undoubtedly more co-ordinated through a particular major provider, that in England comes via a mix of separate contracts for different types of environmental objective and/or different agri-policy instruments. Regardless of these differences, it can be concluded that soft measures appear most effective when they focus on promoting to farmers a common set of practical approaches, derived from and delivered alongside specific environmental insights, which are close to the business motivations of farmers but able to go beyond their immediate business concerns and perspectives, and which are well-informed by both environmental expertise and community understanding. These characteristics enable farmers to appreciate the environmental goals being pursued and the efficacy of related action, to apply this information to their own individual situation, to be supported in making environmentally-appropriate changes in that context, and to reflect and learn from the result of these changes in ways which act as a stimulus to further environmentally-beneficial actions.

Achieving environmental improvements has tended to settle on the common denominator of improved profitability through resource efficiency or no/low-cost amenity actions, which, whilst important on its own, is insufficient to address societal demand. Nevertheless, the most dynamic facilitated farmer groups and local integrated partnerships appear to be taking on a more ambitious set of future challenges (e.g. soil management, climate change adaptation and mitigation) with positive input from government and other environmental stakeholders, which should increase their potential for significant and lasting impact, in future.

Policy makers should pay increased attention to the barriers which prevent or discourage farmers from taking up the advice, training and information offered by private and public services, and design targeted approaches which can overcome these barriers, learning from existing best practice and being sensitive to the context(s) within which they will operate. In particular, it is recommended that policies and approaches using soft measures should adopt socially-embedded, "community of learning" and interactive delivery methods, as these appear to have more lasting impacts than more formal, uni-directional or impersonal methods.

Farm Advisory Services in Greece

Key messages

- Notwithstanding policy makers' recognition of the importance of farmers' advisory, training and extension services to improve the agricultural sector's competitiveness and safeguarding the quality of resources in rural areas, Greek farmers abstain from vocational training and information actions oriented to the protection of the environment.
- The eligibility criteria for the provision of the farm advisory services appear to be biased against small-scale, high-value agriculture and to favour larger holdings involved in the production of arable crops. Only a very small number of farmers benefit from the Farm Advisory System (FAS) (0.3% of farmers who receive direct payments).
- A qualitative assessment of economic performance of the FAS led to negative results due to the limited uptake (benefits), as compared to relatively high start-up and operational costs, although overall cost appears modest in absolute terms. Cost effectiveness is aggravated by the relatively costly one-to-one delivery of advice approach that has prevailed. In general, benefits are insignificant and disproportionate to the costs incurred.
- Greater flexibility of the FAS is needed so that specific and targeted advice is available to farmers. Moreover, agri-environmental advice should be integrated with advice related to productivity, economic- and sustainability-related issues. Monitoring, which is currently confined to observing uptake of advisory services, should be more comprehensive and include evaluations of advisory activities.

Description

The importance of farmers' advisory, training and extension services in enhancing the agricultural sector's competitiveness and safeguarding the quality of resources in rural areas is stressed in the National Plan of Rural Development Strategy (NPRDS) 2007-13 (Ministry of Rural Development and Food, 2012). Despite this recognition, agricultural advice and vocational training is delivered to farmers in Greece at low pace. Statistical documentation indicates that farmers in Greece abstain from vocational training and information actions oriented toward the protection of the environment.

Figure 5.1. Training farm managers by type of training in selected OECD countries, 2010

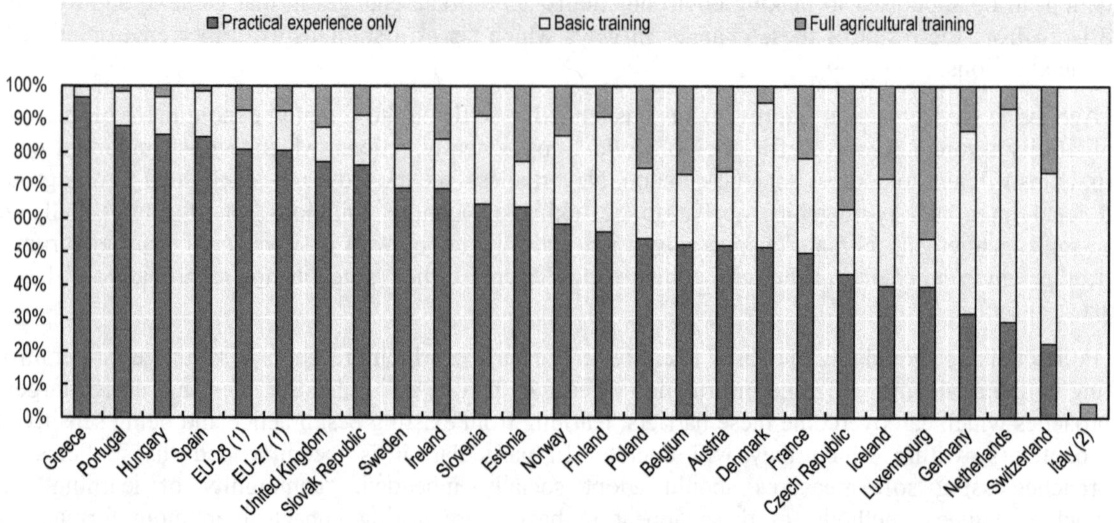

1. Excluding Italy.

2. In Italy different definitions have been used for the levels of "practical experience only" and "basic training". Data from Italy on these categories is therefore not comparable with data from other countries and are not displayed for this reason.

Source: Eurostat (2013), Agri-environmental indicator – farmers' training and environmental farm advisory services, http://epp.eurostat.ec.europa.eu/statistics_explained/index.php/Agri-environmental_indicator_-_farmers%E2%80%99_training_and_environmental_farm_advisory_services.

Figure 5.2. Share of farm managers with vocational training in the last 12 months in selected OECD countries, 2010

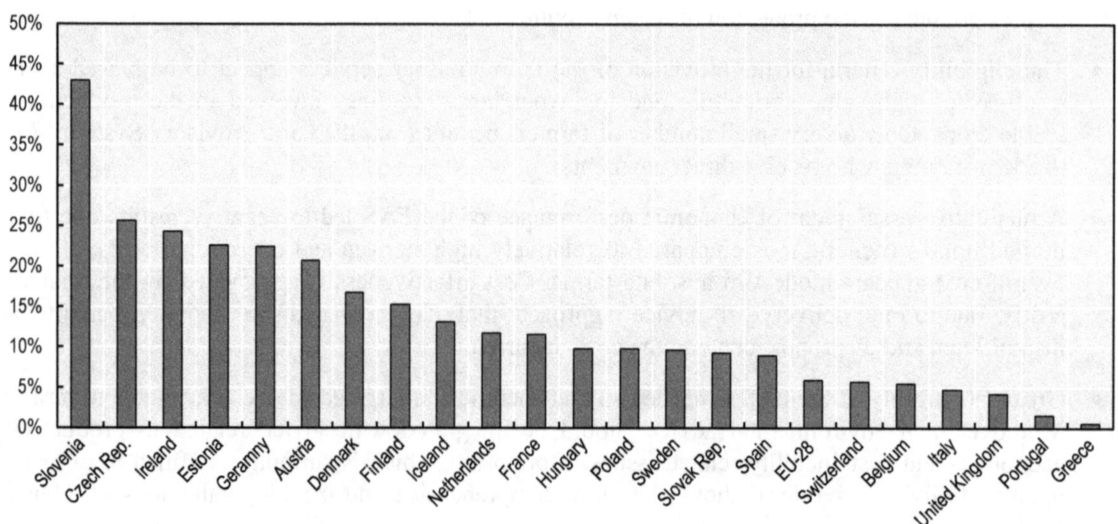

Source: Eurostat (2013), Agri-environmental indicator – farmers' training and environmental farm advisory services, http://epp.eurostat.ec.europa.eu/statistics_explained/index.php/Agri-environmental_indicator_-_farmers%E2%80%99_training_and_environmental_farm_advisory_services.

Farm structure survey data shows that 97% of farmer managers had only practical experience (the highest share in the EU-27, along with Romania and Bulgaria) and occupy 91% (45% for the EU-28) of the total Utilized Agricultural Area (UAA) (Figure 5.1). While at the level of the EU-28, on average, 6% of all farmers have undergone through some process of vocational training in the last 12 months the percentage for Greece is only 0.6% (Figure 5.2). Farms run by managers with basic agricultural training account for 8% of the total UAA (27% for the EU-28) and only 1% (29% for the EU-28) of the total UAA is left to holdings whose managers have full agricultural training. The figures for the EU-28 are 45%, 27% and 29% respectively. Fewer than 2 500 out of a total number of over 700 000 farm managers had full agricultural training as their highest training level.

In compliance with the 2003 CAP reform, a Farm Advisory System (FAS) was set up in Greece in 2007. It was perceived and organised in relation to the NPRDS for the 2007-13 period with the sole aim of complying with the cross-compliance requirements – it has been designed to be independent of the previous extension system, which has become practically obsolete. The delivery of advisory services was assigned to private advisors, whereas the state maintained the role of the regulator.

It was decided, among other things, that: farmers would be able to receive advice services from the advisor of their choice; support for the use of advisory services can be provided up to 80% of the eligible cost per advisory service, with a maximum support amount of EUR 1 500 per farmer in total; Only charges of over EUR 250 per advice segment would be eligible for support; priority was to be given to farmers receiving over EUR 15 000 of direct payments per year, to farms in Natura 2000 and nitrate-prone areas (i.e. areas cultivated with cotton and tobacco); and the one-to-one, on farm advice delivery approach was judged as the most appropriate. The FAS in Greece was partly funded by the European Agricultural Fund for Rural Development (EAFRD) and farmers were expected to contribute an amount equal to 20% of the cost of advice received.

The initial national goal was to accommodate 25 000 farmers and 5 000 foresters by means of implementing the NPRDS's measure 114 (use of advisory services) under the *Improving the competitiveness of the agricultural and forestry sector axis*. This over-optimistic target was subsequently revised to 3 859 farmers and no foresters – and the call for tender, late in the year 2008, led to the approval of 3 882 applications, a figure which exceeds the revised target by a small number of farmers. Many farmers whose applications were approved decided to deny participation in advice delivery, hence the final list was reduced to only 2 148 farmers.

The accreditation process lasted until May 2011, and resulted in the appointment of to 31 operating bodies and 1 055 farm advisers. There is one advisor available for every 862 farm holdings that receive direct payments (909 230 holdings). For holdings that receive more than EUR 10 000 per year (130 950 holdings) there is one advisor available for every 124 farms. The number of farm holdings corresponding to an advisor (regardless of the time each advisor allocates on FAS with respect to other duties) is over double the EU average (ADE, 2009).

Evaluation

The FAS system operating in Greece can be characterised as ineffective primarily due to the miniscule uptake observed given the relatively large number of farmers who receive direct payments. In particular, less than 0.3% of all farmers (about 2 200 farmers) who receive direct payments meet the threshold of receiving direct payments of more than EUR 15 000. This very low uptake performance is compared with an EU-27 average of 4-5%. Very limited advisory services uptake restricts any indirect positive effects with respect to a more efficient use of natural resources. Unfortunately, the quality of advisory services provided cannot be easily assessed due to lack of available sources.

The 2 000 holdings that were offered access to services are, in general, larger commercial farms specialising in the production of specific commodities. Furthermore, no convincing evidence is available that the degree of interaction between farm households and advisory services, as ascertained by the extent that the desired innovative behaviour is adopted by farmers, is significant. It is far from certain that even the small number of beneficiaries has successfully adopted the adjustments needed for securing environmental sustainability.

Opportunities for advice delivery were seriously constrained by a very low demand for such services. Any possibilities offered by the farm advisory system currently operating at the EU level have been under-utilised in Greece, as farmers' willingness to participate, under the prevailing circumstances, has been extremely limited. The approach designed and promoted by the authorities, especially with respect to the selection criteria and the obligations imposed upon farmers, was not attractive to them. The selection criteria and the weights attached to each one reveal that small-scale, high-value agriculture was disregarded and emphasis was placed on larger holdings involved in the production of arable crops.

An empirical study that attempts to assess farmers' attitudes toward educational and training programmes, conducted in northern Greece, concludes that older and uneducated farmers have a negative attitude towards agricultural education and training (Charatsari et al., 2011). On the other hand, educated and younger producers appear willing to spend money and devote time in order to participate in agricultural training programmes depending on the trainers' motivation, competence and effectiveness, the programmes' contents and service delivery techniques. Willing farmers seem to favour training and advice delivery modules that give emphasis to sustainable agricultural practices, as well as the proper application of agri-environmental measures.

Another empirical investigation carried out in the region of Thessaly, in central Greece, that aimed at determining the agricultural training needs of livestock producers, revealed that farmers' perceived training needs were affected by age, education level and farm size. The livestock farmers surveyed demonstrated low levels of confidence in farm advisors. They considered that such programmes were inconsistent with their specific needs. In general, farmers proved to be unwilling to participate in educational and training programmes (Lioutas et al., 2010).

Methods of control, monitoring and imposition of penalties are uniform among all EU member states. However, farmers' awareness, knowledge, compliance and general attitude vary throughout the EU (Diakosavvas, 2011). Many member states are facing implementation difficulties, as is indicated by the penalties being imposed. Among them, Greece is faced with the challenge of providing effective policy advice. Farmers are more likely to comply if they acquire sufficient knowledge and understanding of the relevant obligations (Alliance Environment, 2007). Farmers' compliance is hindered by major shortcomings in knowledge and understanding of the relevant obligations – a fact that provides convincing evidence of policy ineffectiveness in promoting sustainable agriculture.

No official assessment has been made, up to date, about the systems' effectiveness in raising farmers' awareness about the impact certain farming practices have on the environment. On the basis of the nature of the advisory services offered, it can only be assumed that beneficiaries were effectively assisted in meeting cross-compliance requirements. It is doubtful, however, whether the FAS has been instrumental in improving farmers' perceptions of the agri-environmental dimension of policies for agriculture and rural areas. There can be no optimism that the system, as it currently stands, can become more effective in Greece as it is inadequate in addressing the needs of a substantial percentage of the very large farming community.

Lessons learned and policy recommendations

Ensuring that all farmers in the country have equal access to adequate advisory services is vital

Ensuring that all groups of farmers in the country have *equal access* to adequate advisory services is an important issue. Effective promotional activities should be used to provide farmers with in-depth information on the FAS and the potential benefits it has to offer. Small farms in Greece, in spite of their modest contribution to agricultural production in terms of volume, account for a significant part of the family-farm population and, most importantly, they play an important role from an agri-environmental point of view, in terms of land use (ADE et al., 2009; EC, 2010b). Besides, in a situation of economic crisis, small farms can play a supportive role in maintaining employment and income, thus contributing to agriculture, rural development and, potentially, environmental sustainability.

Information campaigns should also encompass small farmers, who need to be convinced about the merits of conforming to sustainability and cross-compliance requirements.

Advice should be tailored to farmers' individual characteristics and preferences

Adjusting agri-environmental advice to satisfy the needs of small farms would further encourage uptake. In defining effective and efficient methods of agri-environmental advice delivery, differences in farmers' individual characteristics and preferences should be taken into consideration. Introducing more flexibility into the system, thus making specific and targeted advice available to farmers more often and according to actual needs without exceeding the initial budget requirements, would, in essence, make the measure more efficient and effective. The demands of producers must be well defined, so that advice can be accurately targeted. Well-tailored advice presupposes a clear specification of farmers' specific demands.

Because of the low level of education and training of farmers engaged in conventional production practices, Greek agriculture is in need of a knowledge delivery system suitable to assist a new generation of farmers in their pursuit of a production model that is viable in economic, environmental and social terms.

The one-to-one advice approach best serves the goal of dealing with the specific concerns of participating farms. If obstacles that limit uptake, such as restrictions on the numbers of farmers eligible for advice are removed and participation is significantly enhanced, targeted (thematic) small group advice on the farm (on-farm, one-to-group) addressed to farm enterprises with similar characteristics might be a cost-effective option to consider (EC, 2010c). The discussion group model adopted in Ireland as an integral part of the Advisory Business Plan and as a key activity of green technology transfer in agriculture could be a proper policy for mobilising farmers and enhancing their participation (O'Loughlin, 2012).

Need to develop a more comprehensive monitoring and evaluation system for FAS activities

Monitoring is currently confined to observing uptake. A national system for a thorough monitoring and evaluation of FAS activities needs to be developed. Increasing effectiveness and efficiency would require the design of a monitoring and assessment system aimed at evaluating the impact of the Greek AKIS on national agricultural policy objectives to make sure there can be some feedback on the design of the existing system. The failure of the implementation of FAS in Greece can be seen as a motivation for the authorities to review and overhaul the entire national farm AKIS system. For such an endeavour, aggregated data on the results of monitoring and control would be useful for adjusting the system and training the advisors.

Enhance farmers' perceptions about the importance of the FAS by linking sustainability purposes and productivity enhancement

The low uptake of farm advisory services raises two issues: i) enforcement of cross compliance requirements is not easy; and ii) farmers, in adopting cross compliance, are cautious and reserved as they seem to link advice delivery with inspection, control and sanctions. Hence, the trust of farmers in the advisors and in the FAS system is undermined. Farmers need to be further convinced that adhering to cross-compliance requirements is important not only for sustainability purposes, but also for productivity enhancement and the competitiveness of their farms.

Agri-environmental advice should be integrated with advice related to productivity, economic, climate change, market and sustainability related issues

Although cross-compliance requirements overlap with practices that curtail agriculture's effects on the climate, fostering green growth would be better-served if distinct actions that confront climate change were to be explicitly incorporated in farm advice delivery. Efforts to increase farmers' demand for advisory services could include a widening of the advice spectrum offered by the national FAS system, with the integration of agri-environmental advice with advice relating to climate change, implementation of rural development measures, business plans design, new technologies, market and sustainability-related issues. A much wider advice domain, going beyond cross compliance would evoke farmers' interest, dissolve their suspicion and induce them to adapt and apply agri-environmental measures.

Advisory services should go beyond technical skills and production practices to include rational and sustainable resource use, marketing, financial and business planning. The new European *Innovation Partnership* (EIP) policy can be utilised as a tool for promoting a competitive and sustainable agriculture in harmony with the environment. Jointly delivered advice that comes as a result of partnership work among stakeholders may appear as more attractive to Greek farmers.

The New Zealand Sustainable Farming Fund

Key messages

- The Sustainable Farming Fund (SFF), which provides funds for community-led research projects, is one of the main funding sources for agri-environmental innovation and research in New Zealand.
- Quantification of the overall impact the SFF is not available. The large degree of diversity amongst funded projects and a lack of comprehensive outcome metrics, coupled with methodological difficulties, impede rigorous evaluation of the SFF's overall impacts.
- Evaluations and assessments of the SFF have been based on mixed methods including economic, multiple criteria (environmental, social and economic) and multiple sources of data (e.g. surveys, case studies, project data, previous evaluations and focus groups comparing SFF with other funds).
- Successive evaluations of the SFF indicate that the SFF is good value for money and has been successful in generating a large diversity of farmer-led research projects.
- A key challenge for the SFF will be to develop further mechanisms and processes for post-project extension to ensure that the potential benefits of its portfolio of projects are maximised through encouraging adoption and further innovation.

Description

Established in 2000, the SFF provides funds for community-led research projects with the aim to create economic, environmental and social benefits to New Zealand's primary industries and rural communities. The SFF, which is one of the main funding sources for agri-environmental innovation and research in the country, is administered by the Ministry for Primary Industries.

SFF projects are all about collaboration and farmer[3] groups have ownership of the project from conception through to adoption. One of the SFF's main strengths is that each project must have an identified "community of interest", a group of people from different backgrounds or organisations, who come together to address a common issue. This "community of interest" is usually led with the support of a farmers' group organisation, agribusiness, researcher and/or consultant. The approach of bringing together farmers, scientists, regional councils and other stakeholders has proved very successful as a medium for sharing information between experts and farmers (Nimmo-Bell, 2009).

The fund was originally confined to land-based industries, but was expanded to include aquaculture in 2011. Since 2012/13, a specific funding round has also been conducted for projects concerned with Māori agribusiness. Since its inception, the SFF has contributed NZD 122.7 million of funding to over 900 projects (up to and including the 2013/14 funding rounds).

The SFF can often act as the seed funder to attract further funding (Oakden et al., 2014). This joint funding model can be successful in stimulating projects that market forces and private sector are unable or unwilling to initiate, whilst funding larger and more numerous projects than the public sector alone is able to support.

The SFF attracts a large number of applicants and is significantly over subscribed each year (Oakden et al., 2014). In 2013/14, applications to the SFF were oversubscribed by 2.3 times.[4] This means that the majority of applicants were unsuccessful in their application. In total, 628 different applicants have been successful in securing SFF funding from 2000/01 to 2013/14. The SFF funds projects of varying sizes and costs. The SFF limits funding for an individual project to NZD 200 000 per year of its duration. The average amount of project funding ranges from NZD 100 000 to NZD 150 000.

SFF projects range in duration from a few months to over four years. It should be noted that SFF projects have a maximum three-year term and those projects that extend beyond three years require contract variations. The most common project duration is for the maximum three years, which accounts for over a third of funded projects. Almost a quarter of projects however run for a year or less, while 95% of projects are conducted within four years.

SFF projects have been carried out across all regions of New Zealand as far as the remote Chatham Islands. Some projects (338) have been conducted at a national level. A total of 124 projects were also conducted in two or more regions. Nearly half (413 projects; 46% of all projects) however are aligned to a single area or region.

SFF projects are classified under 24 topic headings and a project can have multiple topic classifications. Almost half of projects (447 projects) however are classified as contributing to a single topic. By far the largest single topic involves projects relating to pests and diseases (105 projects list pests and disease as the sole topic of the project). Pests and disease is also the most common topic overall, with over 20% (189) of all projects listing it as one of the project topics. Decision management, farm management, crop management and technology transfer are also common project topics, although these are often in conjunction with other topics, and could represent adoption of research, technology and methods. Nutrient management and water quality are a common topic combination, and are the environmental focus of many of New Zealand's primary sectors (particularly dairy).

Projects may be aligned to multiple sectors with 16% of projects citing two or more target sectors. The majority of projects (84%), however, have been aligned to a single sector. Dairy has the most number of SFF projects, consistent with the sector's standing as New Zealand's largest export sector. The combined horticultural sector comes second.

Evaluation

It is difficult to gain an accurate appreciation of the overall impact the SFF has had. The large degree of diversity amongst funded projects and a lack of comprehensive outcome data, coupled with methodological difficulties, make an accurate impact assessment of the SFF impossible. This is a drawback of a community co-ordinated agri-environmental response, where, in the absence of logical assessment framework, impacts can often be ascribed to a large number of variables.

Evaluations and assessments of the SFF have instead relied upon surveys and case studies. Overall however, assessments indicate that SFF projects have been successful in delivering significant benefits and have mostly achieved their targeted outcomes (Nimmo-Bell, 2009; Oakden et al., 2014).

As part of the 2014 evaluation of the SFF, the Kinnect Group used a project managers' perspective to ascertain the extent of SFF's contribution to the economic, environmental and social benefit of the New Zealand primary sector (Oakden et al., 2014; Allen, King and Oakden, 2014). The Kinnect Group evaluation surveyed 136 project managers, which in turn covered around 400 projects, interrogating the

SFF database and conducting three case studies that explored the cumulative benefits of a series of interrelated projects in three sectors. The evaluation was undertaken between March and October 2013. The project managers surveyed were ask if the contribution of the SFF was high, considerable, moderate, limited or not at all in relation to export earnings, productivity, profitability and economic benefit.[5,6] The Kinnect Group deemed that a response of "moderate, considerable or high" represented a belief that the SFF had resulted in a positive contribution, and this was one of the strands of evidence they used to assess value overall.

Results from the survey indicate that actual and potential socio-economic benefits of the SFF were significant. Nearly all (96%) SFF project managers believed that the SFF funded projects they have been involved in would be considered to have delivered good "value for money". Project managers also felt that the majority (94%) of projects had achieved their intended results. An earlier assessment of the SFF also concluded that overall, funded projects have been successful in delivering significant benefits and have most likely achieved their targeted outcomes (Nimmo-Bell, 2009).

Over half of project managers believed that their SFF projects had preserved productivity from threats (helped to protect, safeguard or maintain productivity). A similar proportion (52%) reported that SFF projects contributed to productivity gains at the project level, while about 49% believed that their projects had dispersed productivity gains at a wider regional or sector level. Most productivity gains translated to profitability as nearly half (49%) of surveyed project managers believed that their SFF projects had contributed to a profitability increase at the project level and about 40% percent believed their projects had produced increased profitability at a wider regional or sector level (Oakden et al., 2014).

The SFF contributed to the competitiveness of the primary sector through knowledge creation of successful farming/growing techniques. In terms of farmer knowledge, 88% of SFF project managers believed that farmer participants were now better informed about new and successful farming or growing techniques. In terms of actually applying these techniques, 81% believed that participant farmers were now better able to apply the techniques developed and disseminated. The majority (80%) of SFF managers also believed that the projects assisted in the actual application of these techniques after they had been implemented (Oakden et al., 2014). 43% of project managers believe that SFF projects contribute to New Zealand's "green' credentials", which can be leveraged in international markets.

The SFF has demonstrated success in strengthening the environmental practices of farmers. The survey found that two thirds (66%) believed that their projects had improved environmental practices amongst project participants. Beyond project participants, over half (54%) of managers believed that SFF projects had contributed to improved environmental standards or behaviour at an aggregate agricultural sector or primary industry level (Oakden et al., 2014). This sector/industry impact would typically involve both agro-food industry and private sector (corporate) agri-environmental initiatives. These perceived environmental gains appear to come along with perceived economic gains

In terms of project selection, there was a perception that the SFF may be overly risk averse. Given SFF's stated intention to support grass-roots innovation and the high success rate of projects to date, Kinnect Group (Oakden et al., 2014) recommended that MPI might want to consider whether its application process or risk assessment and selection criteria was excluding some promising but riskier projects, particularly those for Māori.

Measuring social effects, the SFF has been successful in building, applying and transferring knowledge and skills. The majority of surveyed project managers (91%) said that the SFF projects had built the teams' knowledge and skills (Oakden et al., 2014). This degree of influence indicates that groups or individuals outside the projects' original "community-of-interest" can often adopt the knowledge outputs of SFF projects.

Oakden et al. (2014) identified a key social outcome – that SFF projects not only support the building of knowledge, but also the transfer to other communities and stakeholders. Being SFF funded gives a project a degree of credibility amongst the wider community. Project groups discuss their findings with other farmers and stakeholders, often at organised gatherings co-ordinated through industry groups. SFF findings may also be converted into, or contribute towards, management tools that can be adopted by a wider audience than the original project participants.

In assessing the SFF's "overall value for money", Oakden et al. (2014) used multiple criteria (environmental, social, cultural and economic) and multiple sources of data. Projects with a high 'value for money' were defined as having clear evidence of a positive return on the SFF investment together with a credible contribution to export opportunities and improved sector productivity, increased environmentally sustainable practice, and contributions to enhanced environmental, social and cultural outcomes. Although a positive return on investment alone would represent a low threshold, these criteria taken together represent a high threshold for value for money. Similarly, definitions were provided for considerable and moderate value for money (Oakden et al., 2014).

Lessons learned and policy recommendations

The analysis of the project database indicates that the SFF has been successful in generating a large diversity of farmer-led research projects. SFF funded research has ranged from small scale and targeted projects that focused on a particular region, sector or topic, to wider larger scale research that benefited an entire sector or region incorporating multiple topics. The most common profile of an SFF project is one that receives less than NZD 50 000 in SFF funding, conducted over a timeframe of two to three years, and focussed on a single topic and sector. While the majority of applicants and projects are one-off endeavours, there is scope within the SFF for groups to make multiple applications and for projects to flow through into new research.

Available data does not support economic valuation of the impacts of the SFF as a whole due to a lack of logical assessment framework and comprehensive output metrics and the large degree of diversity amongst funded projects. A 2014 evaluation of the SFF involved three case studies and a survey of 136 SFF project managers (responsible for approximately 400 projects) to ascertain the extent of SFF's contribution to the economic, environmental and social benefit of the New Zealand primary sector. Results from the survey yielded a consistent belief from 46% of project managers that the SFF either protected or grew economic value. Over half of project managers believed their SFF projects had preserved productivity from threats (helped to protect, safeguard or maintain productivity).

The SFF contributed to the competitiveness of the primary sector through knowledge creation of successful farming/growing techniques as well as contributing to New Zealand's "green" credentials. The SFF has demonstrated success in strengthening the environmental practices of farmers with two thirds of survey respondents believing that their projects had improved environmental practices amongst project participants. The SFF generates a number of social benefits including increased dialogue between stakeholders and increased use of science to overcome farming obstacles. SFF projects not only support the building of knowledge, but also the transfer to other communities and stakeholders.

The cost-effectiveness of the SFF stems from two different angles. The first involves the efficiency in which the fund is administered; the process of funding application and project selection, the types of projects selected and leveraging upon previous and current research. The second angle concerns the maximisation of the desired outcomes and impacts. This includes ensuring that objectives are clearly defined and measurable, that these objectives are achieved and reported, findings are disseminated, and evaluation is conducted to ensure that projects and the SFF as a whole is having the desired continuing impact.

In terms of administrative efficiency, the SFF has the flexibility to extend project timeframes to strengthen ability to deliver on objectives and the ability to exploit synergies amongst "clusters" of projects. But the SFF suffers from inability to scale-out or scale-up projects after initial SFF approval, a conservative risk approach and significant opportunity cost linked to funding oversubscription. In terms of improving outcome and impact effectiveness and evaluation, there is a need for multi-dimensional end-of-project and post-project reporting as well as wider and more effective dissemination of findings.

A key challenge ahead for the SFF will be to develop further mechanisms and processes for post-project extension to ensure that the potential benefits of its portfolio of projects are maximised through encouraging adoption and further innovation. In order to improve outcome and impact effectiveness and evaluation, there is a need for multi-dimensional end-of-project and post-project reporting.

Despite the positive assessment of the SFF impact on productivity and profitability mentioned above, further analysis of Kinnect Group's series of survey questions on economic outcomes revealed that a quarter of project managers did not know the economic outcomes of their projects, or felt that economic outcomes were not applicable to their projects. This reveals a potential sustainability gap for some farmers (i.e. not addressing one of three SFF sustainability criteria involving economic, social and environmental dimensions) when economic outcomes are not considered and evaluated at the project level. This may also imply that environmental and social values that are hard to monetise far outweigh economic values.

Respondents from the project manager survey also commented that there is an opportunity for Ministry for Primary Industries to assist in the dissemination of SFF project findings and recommendations, providing the role of a centralised hub to proactively ensure the learnings flow across project and sector boundaries. There was an overall feeling that Ministry for Primary Industries needed to more effectively communicate project findings to the wider community.

Finally, the existing assessment studies have several methodological limitations which weaken their ability to make a rigorous assessment of the overall impact of the SFF. For example, the project managers' responses to questions regarding their belief in the contributions their projects had made provide only a general indication that the SFF has a credible prospect of providing a positive return on investment, but this impression cannot be directly substantiated with available data. A parallel survey of "communities of interest" would have validated the views of projects managers who might have an inherent positivity bias to report good results owing to SFF as a funding source for multiple projects.

Notes

1. The *Prime Vert* programme is the main agri-environmental programme in Quebec. It provides financial from the construction of manure storage facilities to support for agri-environmental advisory clubs.
2. http://www.feedadviserregister.org.uk/.
3. The words "farmer" and "farmers" refers to farmers, growers, foresters and aquaculturalists, and business owners and managers.
4. Oversubscription refers to projects applying for funding compared with amount granted and is inn terms of project value. The ratio of number of applicants and projects relative to actual number of successful applicants and projects is not available.
5. It is of interest to note that there is no question on environmental impacts.
6. No information whether projects were randomly selected and period covered (e.g. projects from 2001 or recently completed projects from 2009).

Bibliography

Agriculture and Agri-Food Canada (AAFC) (2013), *Growing Forward 2: A federal - provincial – territorial Framework agreement on agriculture, agri-food and Agri-based products policy*, www.agr.gc.ca/eng/about-us/key-departmental-initiatives/growing-forward-2/?id=1294780620963.

AAFC (2008), *Growing Forward: A federal - provincial – territorial Framework agreement on agriculture, agri-food and Agri-based products policy*, http://repiica.iica.int/docs/B2994I/B2994I.PDF.

ADE (Aide à la décision économique) (2009), *Evaluation of the Implementation of the Farm Advisory System.* ADE in collaboration with ADAS, Agrotec and Evaluators EU. Belgium. http://ec.europa.eu/agriculture/eval/reports/fas/index_en.htm.

Acil-Tasman (2010), *Impact of Investment in Research and Development by the Rural Research and Development Corporations*, Acil-Tasman Pty. Ltd. Canberra, www.ruralrdc.com.au.

Acil-Tasman (2008), *Measuring Economic, Environmental and Social Returns from Rural Research and Development Corporations' Investment*, Acil-Tasman Pty. Ltd. Canberra, www.ruralrdc.com.au.

Allen (2012), *Evaluation of the National Primary Industries RD&E Framework*, Allen Consulting Group, Australia, http://www.npirdef.org/cms news/article/1/9.

Allen, W., J. King and J. Oakden (2014), *Three case studies: companion document to the Evaluation of the Sustainable Farming Fund*, prepared for Ministry for Primary Industries, Kinnect Group, Wellington.

Alliance Environment (2007), *Evaluation of the application of cross compliance as foreseen under Regulation 1782/2003, Part I: Descriptive Report – Final*, http://ec.europa.eu/agriculture/eval/reports/cross_compliance/full_text_en.pdf.

Alston, J., M. Andersen, J. James and P. Pardey (2011), "The Economic Returns to U.S. Public Agricultural Research", *American Journal of Agricultural Economics*, Vol. 93, No. 5.

Bell, B. and M. Yap (2015), "The Performance of the Sustainable Farming Fund in New Zealand", an OECD internal document.

Birner, R, K. Davis et al. (2006), *From "best practice" to "best fit": a framework for designing and analyzing pluralistic agricultural advisory services worldwide,* EPTD Discussion Papers 05, International Food Policy Research Institute (IFPRI). Washington, DC.

Charatsary, C., A. Papadaki-Klaudianou and A. Michailidis (2011), "Farmers as consumers of agricultural education services: willingness to pay and spend time", *Journal of Agricultural Education and Extension*, Vol. 17, No. 3.

Cuevas-Cubria, C., C. Gibbs et al. (2012), *Measuring and reporting trends relating to the performance of Australia's rural RD&E system*, ABARES report to client prepared for the Agricultural Productivity Division, Department of Agriculture, Fisheries and Forestry, Canberra, June.

Damianos, D. (2015), "An Analysis of the Performance of Farm Advisory Services as related to Fostering Green Growth – The Case of Greece", Consultant Report, OECD internal document.

Department for Environment, Food and Rural Affairs (DEFRA) (2013), *Review of environmental advice, incentives and partnership approaches for the farming sector in England*, www.gov.uk/government/publications/review-of-environmental-advice-incentives-and-partnership-approaches-for-the-farming-sector-in-england.

DEFRA (2010), *Agricultural Advisory Services Analysis*, http://archive.defra.gov.uk/foodfarm/landmanage/climate/documents/advisory-analysis.pdf.

DEFRA (2008), *Environmental Stewardship Review of Progress, Defra and Natural England.* Department for the Environment, Food and Rural Affairs, London. http://collections.europarchive.org/tna/20081027092120/http://defra.gov.uk/erdp/schemes/es/es-report.pdf.

Department of Agriculture, Fisheries and Forestry, Commonwealth of Australia (2008), *Making a difference: A celebration of Landcare.* www.daff.gov.au/natural-resources/landcare/publications/making_a_difference_a_celebration_of_landcare.

Dwyer, J. and M. Reed (2015), "Promoting Green Growth in English and Welsh Agriculture: Evaluating the Role of "Soft Measures" – Training, Advice and Extension", Consultant Report, OECD internal document.

Diakosavvas, D. (2011), "Greening Agricultural Policies in OECD Countries: What Role for Cross Compliance?", OECD Trade and Agriculture Directorate, PRIMAFF Seminar, Tokyo.

Ekboir, J. (2003), "Why Impact Analysis Should Not Be Used for Research Evaluation and What the Alternatives are", *Agricultural Systems*, Vol. 78.

European Commission (EC) (2010a), *On the application of the Farm Advisory System as defined in Article 12 and 13 of Council Regulation (EC) No. 73/2009*, Report from the Commission to the European Parliament and the Council, COM(2010) 665 final, http://eur-lex.europa.eu/legal-content/EN/TXT/?uri=CELEX:52010DC0665.

EC (2010b), "Main issues and evaluation of the implementation of the Farm Advisory System (FAS) in Member States", DG Agriculture and Rural Development – Unit D3, FAS Workshop – Barcelona 10 June, http://www.google.fr/url?sa=t&rct=j&q=&esrc=s&frm=1&source=web&cd=1&ved=0CCMQFjAA&url=http%3A%2F%2Fmars.jrc.ec.europa.eu%2Fmars%2Fcontent%2Fdownload%2F1807%2F9794%2Ffile%2Finge_van_oost.pdf&ei=HzIUVZ6lL8XYapKpgsgF&usg=AFQjCNHLogKPBUNuiCY5bgBX8kpXkUO1dA.

EC (2010c), "Report from the Commission to the European Parliament and the Council on the application of the Farm Advisory System as defined in Article 12 and 13 of Council Regulation (EC) No. 73/2009, COM(2010) 665 final", European Commission, Brussels, http://eur-lex.europa.eu/legal-content/EN/TXT/?uri=CELEX:52010DC0665.

Eurostat (2013), *Agr-environmental indicator – farmers' training and environmental farm advisory services*, http://epp.eurostat.ec.europa.eu/statistics_explained/index.php/Agri-environmental_indicator_-_farmers%E2%80%99_training_and_environmental_farm_advisory_services.

Framework (2014), *National Primary Industries RD&E Framework*, http://www.npirdef.org.

Gibbs, C. (2013), "The Value of Advice Report", Agricultural Industries Confederation Ltd., Peterborough, UK.

Grant, A. (2012), "Australia's approach to rural research, development and extension", in OECD, *Improving Agricultural Knowledge and Innovation Systems: OECD Conference Proceedings*, OECD Publishing, Paris.
DOI: http://dx.doi.org/10.1787/9789264167445-6-en.

Gray, E., M. Oss-Emer and Y. Sheng (2014), *Past Reforms and Future Opportunities*, Australian Bureau of Agricultural and Resource Economics and Sciences (ABARES), Research Report 14.2, Canberra.

Kefford, B. and C. Noble (2015), "Australian Primary Industries Research, Development and Extension System – Its Evolution, Features and Impacts on Green Growth Objectives", Consultant Report, OECD internal document.

Laurent, C. and P. Labarthe (2011), "Économie des services et politiques publiques de conseil agricole", *Cahiers Agricoles*, Vol. 20, No. 8 5 September-October, pp. 343-351.

Lioutas, E.D., I. Tzimitra-Kaloyianni and C. Charatsari (2010), "Small Ruminant Producers' Training Needs and Factors Discouraging Participation in Agricultural Education/Training Programs", *Livestock Research for Rural Development*, Vol. 22, No. 7.

MacDonald, N., J. Dwyer et al. (2006), "The influence of produce protocols on water use and land management. EA project A20: Project report to the Environment Agency of England and Wales by ADAS and CCRU", (unpublished), February 2006.

Mills, J., N. Lewis and J. Dwyer (2010), *The Benefits of LEAF Membership: a qualitative study to understand the added value that LEAF brings to its farmer members*, CCRI, Gloucester.

Ministry of Rural Development and Food (2012), *National Strategy Plan 2007-13* (in Greek).

Natural England (2014), *Making Environmental Stewardship More Effective (MESME), Defra – Natural England 2008*, NE, Peterborough.

Nimmo-Bell (2009), *Evaluation of Sustainable Farming Fund Projects, prepared for MAF Sustainable Farming Fund*, Nimmo-Bell & Company Ltd., New Zealand.

Oakden, J., J. King and W. Allen (2014), *Evaluation of the Sustainable Farming Fund: Main Report*, prepared for Ministry for Primary Industries, Kinnect Group, Wellington.

OECD (2015a), *Innovation for Agriculture Productivity and Sustainability: Review of Australian Policies*, OECD Publishing, Paris, forthcoming.

OECD (2015b), *Innovation for Agricultural Productivity and Sustainability: Review of Canadian Policies*, OECD Publishing, Paris, forthcoming.

OECD (2012), *Improving Agricultural Knowledge and Innovation Systems – OECD Conference Proceedings*, OECD Publishing, Paris
DOI: http://dx.doi.org/10.1787/9789264167445-en.

O'Loughlin, L. (2012), "Development of Discussion Groups and Mobilizing Farmer Participation, Teagasc, Best Practice in Extension Services Conference – Supporting Farmer Innovation", Conference Proceedings, 1 November, Dublin, www.teagasc.ie/publications/view_publication.aspx?PublicationID=1590.

Pearce, D. and E. Willliams (eds.) (2010), *Seeds for Change. Action Learning for Innovation,* Menter a Busnes, Aberystwyth.

Prager, K., R. Creaney and A. Lorenzo-Arribas (2014), "Advisory services in the United Kingdom: Exploring "fit for purpose" criteria", *The James Hutton Institute*, Craigiebuckler, Aberdeen, Scotland, UK.

Productivity Commission (2011), *Rural Research and Development Corporations*, Report No.52, Final Inquiry Report, Canberra, www.pc.gov.au/projects/inquiry/rural-research.

Reed, M. and P. Courtney (2013), "Developing farm-level social indicators: Measuring their networks, well-being and capacity for social innovation", paper presented at the XXVth European Society for Rural Sociology (ESRS) Congress, Florence, Italy.

Sheng, Y., E. Gray, J. Mullen and A. Davidson (2011), *Public investment in agricultural R&D and extension: an analysis of the static and dynamic effects on Australian broadacre productivity*, Australian Bureau of Agricultural and Resource Economics and Sciences (ABARES), Canberra.

Statistics Canada (STC) (2011), *Farm Environmental Management Survey*, Ottawa, Canada.

STC (2007), *Farm environmental management survey.* www5.statcan.gc.ca/bsolc/olc-cel/olc-cel?catno=21-021-MIE&lang=eng#formatdisp.

Summers, R., R. Plummer and J. FitzGibbon (2008), "Accounting precautionary principles measures in agriculture trough pathway analysis: the case of the Environmental Farm Plan", *Journal of Agricultural Resources, Governance and Ecology*, Vol. 7, No. 6.

Tchuisseu, R. and P. Labarthe (2015), "Investing in Knowledge to Support the Adoption of Environmentally-friendly Practices: Lessons Learned from the Canadian Growing Forward Program", Consultant Report, OECD internal document.

ORGANISATION FOR ECONOMIC CO-OPERATION AND DEVELOPMENT

The OECD is a unique forum where governments work together to address the economic, social and environmental challenges of globalisation. The OECD is also at the forefront of efforts to understand and to help governments respond to new developments and concerns, such as corporate governance, the information economy and the challenges of an ageing population. The Organisation provides a setting where governments can compare policy experiences, seek answers to common problems, identify good practice and work to co-ordinate domestic and international policies.

The OECD member countries are: Australia, Austria, Belgium, Canada, Chile, the Czech Republic, Denmark, Estonia, Finland, France, Germany, Greece, Hungary, Iceland, Ireland, Israel, Italy, Japan, Korea, Luxembourg, Mexico, the Netherlands, New Zealand, Norway, Poland, Portugal, the Slovak Republic, Slovenia, Spain, Sweden, Switzerland, Turkey, the United Kingdom and the United States. The European Commission takes part in the work of the OECD.

OECD Publishing disseminates widely the results of the Organisation's statistics gathering and research on economic, social and environmental issues, as well as the conventions, guidelines and standards agreed by its members.

OECD PUBLISHING, 2, rue André-Pascal, 75775 PARIS CEDEX 16
(51 2015 04 1 P) ISBN 978-92-64-23215-0 – 2015

www.ingramcontent.com/pod-product-compliance
Lightning Source LLC
LaVergne TN
LVHW081417110826
845149LV00010B/1779

* 9 7 8 9 2 6 4 2 3 2 1 5 0 *